АКАДЕМИЯ НАУК УКРАИНСКОЙ ССР
ЦЕНТРАЛЬНЫЙ РЕСПУБЛИКАНСКИЙ БОТАНИЧЕСКИЙ САД

АЛЛЕЛОПАТИЧЕСКОЕ ПОЧВОУТОМЛЕНИЕ

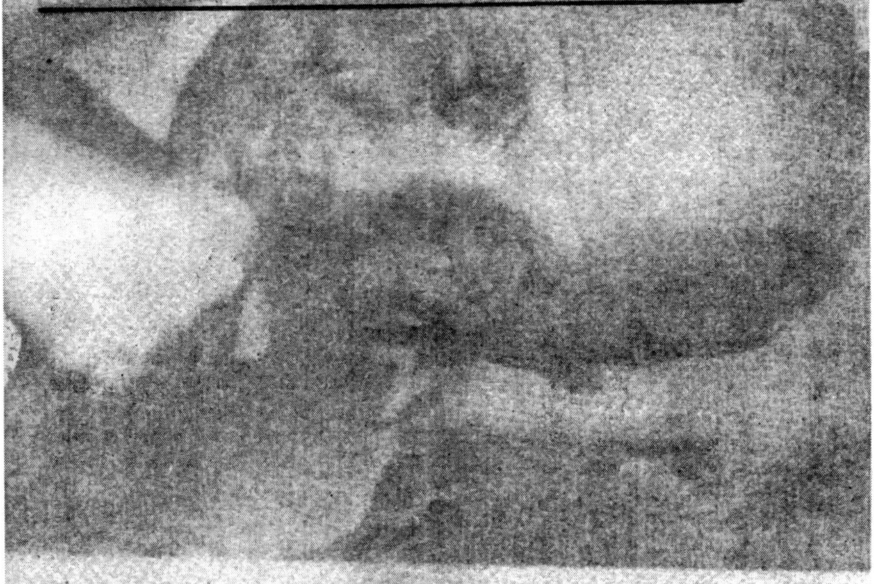

КИЕВ «НАУКОВА ДУМКА»
1979

Professor Andrei Mikhailovich Grodzinsky
(03.12.1928 - 17.12.1988)

Academy of Science of Ukrainian SSR

Central Republican Botanical Garden, Kiev, Ukraine, USSR

ALLELOPATHY IN SOIL SICKNESS

Editor-in-chief

A.M. Grodzinsky

English Editor

S.S. Narwal

Translator

Barbara Polyticka

SCIENTIFIC PUBLISHERS (INDIA), JODHPUR

INTERNATIONAL ALLELOPATHY FOUNDATION, HISAR

Published by:
Scientific Publishers (India)
5-A, New Pali Road, P.O. Box 91
Jodhpur - 342 001 (INDIA)
E-mail: info@scientificpub.com
www.scientificpub.com

Published for
International Allelopathy Foundation
8/15, Haryana Agricultural University
Hisar, 125 004, India
E-mail: allelopathy1947@yahoo.com

Print : 2021

ISBN: 81-7233-423-0

Lasertype set: Rajesh Ojha
Printed in India

EDITOR AND TRANSLATOR

Prof. S. S. Narwal

Prof. Barbara Politycka

Prof. S. S. Narwal is an Internationally known allelopathy scientist. He is the first National Fellow (Allelopathy) of Indian Council of Agricultural Research, New Delhi, since 1995. Currently he is President, Indian Society of Allelopathy and the Chief Editor, Allelopathy Journal.

He has travelled to more than 35 countries of world, to delivers Lectures in Allelopathy Conferences or as visiting Professor. In 1989, when he was Visiting Professor, Department of Allelopathy, Ukraine, Academy of Science, Kiev, he attended Soviet Allelopathy Conference there and was allotted the room of world famous scientist Prof. A. M. Grodzinsky for office work. There, he read thousands of Reprints from Grodzinsky's Collection. This broadened his vision about the future scope of Allelopathy. Prof. Narwal founded Indian Society of Allelopathy in 1990. Prof. E. L. Rice, USA wrote Congratulations Prof. Narwal, it is the first National Society in the World". He has organized 4 International Allelopathy Conferences on Allelopathy in India viz., 1992 (Hisar), 1994 (New Delhi), 1998 (Dharwad), 2004 (Hisar). He is also the founder of International Allelopathy Society (1994) during the II International Allelopathy Conference at New Delhi. He has authored/ edited 15 Books.

Prof. Barbara Politycka is Scientist in the Department of Plant Physiology, August Cieszkowski Agricultural University, Poznan, Poland. She had been conducting Allelopathy research and teaching the Allelopathy to classes in Plant Physiology, since 1973. Her studies are the continuation of the earlier achievements of Department, the direct action of biologically active substances originating from the decay of organic matter in the soil. Her major scientific interests are: (i). Allelopathy in soil sickness phenomenon and (ii). Plant responses to phenolic compounds released from plant residues. She primarily studies, the participation of phenolic compounds in the phenomenon of soil sickness and plant responses to these allelochemicals. The main topics of her studies are:

- Soil sickness of organic substrates used repeatedly in cucumber growing in Glass houses.
- Release of phytotoxic substances from the decaying cereal crop residues.
- Mechanisms of physiological responses of cucumber to phenolic allelochemicals.
- Phenolic compounds as one of the casual factors of apple replant problem.

She is the co-author of a Monograph and numerous publications on Allelopathy. In 2004, she was awarded the 'Life Time Achievements Award, by International Allelopathy Foundation.

FOREWORD

It is a sincere pleasure for me to write a foreword for this Classic Book *Allelopathy in Soil Sickness*. This is the first book on Soil Sickness worldwide in Russian language. It was translated because the English is read in most countries of the world, hence, the allelopathy scientists would be able to know the phenomenon of Soil Sickness. In the past, this problem had been reported mainly from fruit orchards (Apple, Pear, Peach etc) and from some forest trees in temperate countries. However, this problem is now increasing in the tropical and sub-tropical countries due to adoption of Multiple Cropping Systems (e.g. Rice-Wheat, Rice-Rice, Sugarcane Ratoons and Vegetable based Cropping Systems etc.) in several Asian countries. It is hoped that this book will provide inspiration and in-depth knowledge of the complex problem of Soil Sickness.

This book consists of 9 chapters viz., (i). History of soil sickness, (ii). Soil sickness in field crops, (iii). Soil sickness in fruit trees, (iv). Soil sickness in flower cultivation, (v). Chemical nature of soil toxins, (vi). Physiological and biochemical interactions of toxins, (vii). Role of microorganisms in Soil sickness and to restore soil fertility, (viii). Determining allelopathic soil toxicity, (ix). Conclusions and (x). References.

Prof. Grodzinsky was an outstanding Allelopathy scientist, who is called *'Father of Allelopathy'* in USSR. He edited 16 books (1965-1987), highest number of books edited by any other scientist worldwide and authored 100 research papers. He established the first Department of Allelopathy, Central Botanical Garden, Kiev, Ukraine. He worked on Allelopathy for 23 years (1965-1988).

One of the activity of International Allelopathy Foundation, Hisar, India, is to sponsor the translation of classical Books on Allelopathy from Non-English languages to English for the benefit of English scientists. Therefore, since 1995, the International Allelopathy Foundation was trying to get this book translated through various scientists worldwide. Finally, Prof. Barbara Politycka, Poznan, Poland agreed to translate. I am highly thankful to her for sparing her valuable time and spending lot of energy in completing this difficult Job. She has done it in a very lucid way in 3-months. I am also thankful to Mrs. Walentyna Michalska to improve the translated text. I have edited the book for English language and devided the running text in Sections and Sub-Sections to make it more useful to readers.

This book is the second such Translation from the International Allelopathy Foundation. It has already published the English translation of book *Hans Molisch (1937). Allelopathy: Influence of one Plant on Other* in 2002. It is making arrangements for the translation of few more Classic books on Allelopathy to English language, for which the kind cooperation of scientists, who may help in such translation is very much required.

August 22, 2005

Prof. S.S. Narwal
ICAR National Fellow (Allelopathy)
Department of Agronomy,
Haryana Agricultural University
Hisar- 125 004, INDIA

PREFACE

Soil sickness is a loss or a strong decrease in soil fertility as a consequence of growing some agricultural plants or a monoculture of the same plant species. To restore soil fertility, "sick" soil should "rest", i.e. it should be allowed to lie fallow or should be cultivated to reduce soil sickness. Initially soil sickness compelled the farmers to abandon the cultivated soil and to till new plots of virgin steppe to make arable land by cutting the forests. In this way, there were two agricultural systems : long fallow (in steppe) and burning (in forest area). When the farmers came back to the former fields after decades, the soil had a chance to "restore". With the increase in population and the extension of cultivated land areas, the possibilities of very long crop rotations have been reduced and at last a three-field system has been adopted. In this system, the soil rested during only one year and 33% of the entire area in kept fallow. The intensification of agricultural production, the use of machines, chemical fertilizers, pesticides to control crop pests (weeds, diseases and insects) and the introduction of regular crop rotations reduced the fallow area to 10-15%.

However, the existing agriculture keeps survivals of patriarchal "nomadic" farming in secret. The practice shows that the modern agricultural enterprise has only one or two really profitable branches, whereas, the remaining branches create overhead expenses, decreasing the profitability of farmers. Meanwhile, the necessities to keep strictly to crop rotation, farmers are forced to grow many crops. The necessity to practise crop rotation, compels the farmers every year to change abruptly the established system of assigning lands to particular crops. The most ideal alternative is to make the farm specialized in growing monoculture. This opens unlimited possibilities for mechanization and automation of agricultural operations. The development of agriculture in the advanced countries (USA, England, Germany, France) shows greater tendency to a narrow specialization of individual farms, to concentrate on one kind of production, for instance, only maize or green pea or milk or broiler chicken etc.

It may pose a dangerous problem on specialized forms by not following crop rotations. This problem is the soil sickness and the fear to destroy soil fertility forever. Meanwhile specialization in plant production is possible only for few crops, requiring high intensity of tillage. It is true that in different countries including USSR, the experiments on monoculture of winter wheat, rye, potatoes, beet and other plants have been conducted since last more than 130 years and in many cases the yields can be kept high. Nevertheless, the monoculture have seen adopted only for maize, grape, tobacco and some vegetable crops and potatoes. Other crops require crop rotation, soil "rest" and fallow land for their successful growth. The most serious soil sickness is induced by sugar beet, flax, pea, clover, alfalfa, fruit-plant cultures, but in principle, any cultivated plant can make soil sick, if precautions are not taken.

Published papers and reviews on the problem of soil sickness are not too many (Krasilnikov and Garkina, 1946; Grümmer, 1955; Verona, 1963). They show that soil sickness is a complex natural event and has many reasons. Most of these reasons (lack of nutrients, weed and pest multiplication, deterioration of water-physical properties of soil, etc.) can be removed by fertilizers, pesticides or by agrotechniques. However, there are many soil sickness mechanisms connected with the accumulation of physiologically active substances in soil – toxins, which are formed during chemical interactions of plants or allelopathy.

In this book we have used the materials of D.D. Sigareva (WNIS) as well as papers by L.D. Yourtshak, T.P. Bukolova, L.C. Serediuk and other workers of the Allelopathy Laboratory, Central Botanical Garden, Kiev, the Ukrainian Academy of Sciences (AN USSR). We express our deep gratitude for their contributions.

January, 1, 1978

Prof. A.M. Grodzinksy
Department of Allelopathy
Ukraine Academy of Sciences
Central Botanical Garden
1, Timirvzskaya Street
Kiev-202544, Ukraine, USSR

CONTENTS

Chapter 1

SOIL SICKNESS : HISTORY

A.M. Grodzinsky

Department of Allelopathy,
Central Botanical Garden, 1, Timirvzskaya Street, Kiev, Ukraine

"Searching for new fertile lands in ancient Russia was a great event everyday." A.W. Sovyetov, About Systems of Agriculture, 1867 (Sovyetov, 1950)

The phenomen of soil sickness in agriculture is known since ancient times. Long ago, the ancient farmers realised the imposibility of multiple cropping of one plant species on the same field and loss of soil fertility in monoculture. That is one of the reasons of the occurrence and development of different cropping systems.

In the south USSR steppe zone, there is long fallow system of land cultivation. The ploughed land was sown with wheat, rye or barley for many years in succession and after its exhaustion, it was left as long fallow for pasture or haymaking. After 6-7 years more, this land was again sown with crops. In this system, weed control was done by using land as natural steppe and forcing the weeds out, whereas, mineralization of organic substances was achieved by soil tillage and harrowing. Burning of stubbles or steppe grass before the first ploughing was practiced on large scale. Virgil described this method in his poem "The Georgics" in the Ist century B.C. devoted to agriculture (Virgilis, 1971). Tacit wrote about ancient Germans that they changed their fields every year.

Burning: According to Crescenci, rearranging and burning systems were used in the XIIIth century in Europe. For example, the land in Brittany was used for growing wild blackthorn (*Ulex europaeus*) for 7 years, after wards blackthorn was cut down, when its seeds were ripe and thrashed. The drywood was used for burning the blackthorn residues. Then buckwheat was sown with cereals. In the fifth year, oat was sown with blackthorn seeds and this again was followed by 7-years rest. In north, where the land is covered by bushes, the burning and cultivation system was more predominant. In England, *Calluna vulgaris* Salisb. was burnt off each 7-8 years to provide pastures for sheep-breeding. In Russia, the burning system of farming was wide spread long ago. Burning or hew system of agriculture, consisted of burning the forests, bushes and turf to introduce new lands to cultivation. Sowing was done without tillage, simply on ashes.

It was not only the destruction of interfering vegetation (trees, bushes, turf, root offshoots etc. and their primordia), but also the acceleration of mineralization of organic substances available in the biocenosis, to provide mineral nutrition for cultivated plants. It seemed that there was deeper sense in this empirical aspiration not realized so far, because the organic substances in soil, were continuously transformed to and serves as source of kolines – physiologically active substances, which suppress the plant growth, when their concentration was high. The excessive burning in many places led to the destruction of soil fertility. J. Liebig (1861) reported that "*No where consequences of that system were as striking as in America. The first colonists in Canada, New York State, Pennsylvania, Virginia and Maryland territories, with a single ploughing and sowing, got such high yields of wheat grains and tobacco leaves for many years, that the thought to give back what the farmer took from the land could not come to his head. We all know what has happened now with these fields. Succeeding generations did not manage to substitute the nutrients, which the crops are taking from the soil and rich cornfields have converted into a fruitless desert. After the lapse of the entire century, they remained to be a long fallow and in many districts the cultivation of cereal crops is not worth doing*" (J. Liebieg.

Letters about the present state of agriculture, 1861 – cited after Sovyetov, 1950).

With the increase in human population, the possibilities of nomad agriculture had become limited and there appears the system of accelerated soil "rest" called a three-field rotation with alternating fallow, winter and spring crops. This system existed in Europe and Russia for several years and only due to the success of natural sciences and development of new agricultural technology, more complicated crop rotations were introduced into agriculture. The demand for fertilizers was met by the introduced legumes (Norfolk crop rotation, clover sowing), stall manure, bone meal, skin scraps and various waste materials, peat and imported Chilean nitre.

At the times of supremacy of the humus theory of plant nutrition by A. Taer, who suggested that plant builds its body and creates yield at the cost of organic substances or soil humus, wide spread were the ideas about the biologically active root secretions which are poisonous to mother plant. A similar wide spread idea was indicated by a known poem by A. Puszkin "Antiar", in which the poet showed principles of chemical plant interactions, which have appeared to be true now. Root secretions were not toxic to all species – crops causing soil sickness should be replaced by other species, which will use these toxins and improve the situation. However, these species in turn leave new toxins and should be treated in the same way. A similar rotation of species takes place in forest, steppe and in meadow plant communities. It is true that nobody has succeeded to isolate toxins from soil and to show their effect.

Theories of photosynthesis and mineral plant nutrition developed in the mid 19th century, the discovery of nitrogen-fixation by symbiotic and freely living microorganisms, experiments of successful plant cultivation on completely mineral substrate formed the basis of practical agrochemistry. These showed the possibility to overcome the decline in soil fertility by applying mineral fertilizers and mechanized soil tillage.

The cases of soil sickness were sometimes explained by (a) phytopathological diseases, (b) the lack or excess of micronutrients (c) excessive development of pests or (d) improvement in correlation between mineral salts. However, it was not clear why crops with approximately the same requirements of macro- and micronutrients sometimes induce different soil sickness. This peculiarity was probably in the plant nature itself.

At the beginning of the 19th century, studies of the soil sickness started under field conditions (Gortinsky, 1966), in many cases the accumulation of harmful alcohol soluble substances were found in the soil. Pryanishnikov (1976) paid great attention to this problem, considering that the main thing for native soils was the lack of nitrogen and other plant nutrients. Studies of Schreiner (1907-1911), Pikering (1917-1919) and later investigations were discussed in the monograph of Grümmer (1957) and in the review of Verona (1963).

New facts about the accumulation of toxic substances in soil were obtained from the long-term complex investigations of the citrus plants and peach replantation problems in Canada and California (Patrick, 1955). He found that not only the root excretions, but also root residues remaining in the soil after tree cutting determines the stable toxicosis of soil and its influences or the further growth of the same plants. Likewise, ancient scientists have reported the harmfulness of those plants (chickpea, vetch, bean), whose root systems remained in the soil after harvest, because they are not pulled out.

The above reviews attempted to define the term "soil sickness". Gortinsky (1966) understood it as a progressive yield decline due to repeated sowings in the same place. Verona (1963) considered that the lack of development and low productivity of monoculture should be referred to as `soil sickness' independent of soil fertility. Rademacher and his followers (Börner *et al.*, 1959) understood soil sickness as a progressive yield decline, occurring under conditions of repeated plant cultivation on the same soil, which was not influenced by exhaustion of nutrient substances or by other abiotic factors *viz.*, plant diseases and pests. In their opinion, the reason of soil sickness should be searched in the excretions of physiologically active compounds from living plant parts and plant residues, which exert a direct or indirect influence through microflora.

We suggest that soil sickness should be understood more widely. It cannot be limited to cases of monoculture, where this phenomenon is very distinct and most common. It is very easy to foresee a situation, when one crop makes soil sick for another crop (usually called "a bad predecessor") using namely the same ecological mechanism and involving the same substances, which suppress this crop itself. On the contrary, there was data showing a significant difference between the cultivars of one crop in their ability to make soil sick and to tolerate soil sickness. Quite possible is such combination of cultivars – a little active one as a predecessor and a highly tolerant next crop, in which allelopathic soil sickness will not occur.

Soil sickness can have very different reasons, a varying degree of expression and varied stability (depending on the duration of time required for "rest" period). The action of soil sickness intensified with the intensification of cultivation. The most significant reasons for sickness are:

1. One-sided nutrients exhaustion, lack of micronutrients, disturbance in salt balance of soil, "overloading" with fertilizers. In the past, when pesticides containing fluorine, mercury, copper and other elements were widely used, soil contamination with them in isolated places could not be excluded e.g., under fruit-trees.
2. Deterioration of soil structure and its physical and chemical properties, especially in long cultivation of bulb crops.
3. Development of phytopathogenic microflora, intensifying in monoculture.
4. One-sided development of some groups of soil microflora, to the detriment of other groups.
5. Intensifying pest multiplication.
6. Excessive infestation of invasive weeds.
7. Changes in pH.
8. Accumulation of phytotoxic substances in soil.

In a more philosophical view, all scientific efforts were directed to increase the soil fertility and to overcome its sickness. Like in ancient times, the most common method to overcome the soil sickness was the transfer of crop to new place, i.e. crop alternation, shift of crops. Most of the reasons of soil sickness were studied by differenent sciences (Table 1). In case of one-sided development of soil microflora, we put a question-mark because this problem was little studied and it was not known what can be done to improve the balance between microorganisms and what optimal balance should be maintained for the best growth of higher plants. The applied bacterial inoculations did not change the conditions occurring in the soil.

Table 1. Causes of soil sickness and preventive measures

S. No.	Causes of soil sickness	Field of science dealing with soil sickness	Preventive measures
1	Decrease of nutrients content, deficiency of microelements, nutrients imbalance	Agronomy, Plant physiology	Apply different kinds and forms of fertilizers
2	Disturbance in structure and physico-chemical properties of soil changes in pH	Soil science	Cultivation, manuring, liming
3	Development of pathogenic microflora	Phytopathology	Disinfection of seeds, cultivation of disease resistant plants, quarantine
4	One-sided development of some groups of microflora i.e. dominance and reduction of other groups	Soil microbiology	Bacterial inoculation, nitragin azotobacterin, phosphoro-bacterin and others
5	Multiplification of pests	Entomology	Chemical & biological control methods of plant protection
6	Multiplification of invading weeds	Botany, agriculture	Cultivation use herbicides
7	Accumulation of phytotoxic compounds in soil	?	?

The accumulation of phytotoxic substances in soil was mentioned by many scientists (Gortinsky, 1963; Grodzinsky, 1969) but it was very little studied. More publications are devoted to soil sickness under *Citrus* and plants of Rosaceae family fruit-plants (Patrick, 1955; Börner, 1961; Moroz, 1968, 1972; Berestecki, 1970, 1972). The accumulation of unknown toxins, negatively acting on the growth in bioassays or the accumulation of precisely determined substances (such as phloridzin and its derivatives, patulin, saponins etc.), have been found. It is however unknown, what can be done with these toxins and how their contents can be regulated in the soil.

There is no information, how to overcome the soil toxicity. It was suggested that liming of soil may be useful, but is not always effective. The removal of post-harvesting residues or roots when cutting old fruit-trees seems to be advisable, because they are sources of toxins in the soil (Bublitz, 1959; Grodzinsky *et al.* 1962; 1963). The question concerning the accumulation of phytotoxic substances and methods of their removal remains open. It seemed that in natural agriculture, there are existing mechanisms of phytotoxin accumulation, their destruction and removal from the environment, which act without our knowledge. It should be understood that the main principles of crop rotation, selection of the previous and next crops, agrotechnical measures etc. resulted empirically from the centuries-old experience of farmers and it was quite possible, that these rotations remove the soil toxicity. Study of these mechanisms, can undoubtedly lead to increasing soil fertility and agricultural production.

Allelopathy is a rotation of physiologically active substances – kolines (Grodzinsky, 1965, 1974) and is directly related with the problem of soil sickness. Some researchers are of the opinion that allelopathy is generally identical with soil sickness (Proebsting, 1950; Koch, 1955; Börner, 1960; Grümmer, 1970). This was incorrect, since allelopathy not only suppresses the plants, but also stimulates their growth. However, there were many complex factors, related to soil sickness (formation and excretion of kolines and their absorption by other plants, physical and biochemical factors, inter-root metabolite exchange etc.).

If soil sickness is considered as a certain cyclic process, accelerated by the above causes and also having a cyclic character, then the existing relationships can be presented as a scheme (Fig. 1), where these processes are shown as a system of contiguous discs rotating in different directions. The circle marked "soil sickness" is contiguous to eight circles rotating in one direction and presenting causes, which jointly "disperse" soil sickness. There is no direct connection between them – the circles can tow, rotate idle, if "soil sickness" is inhibited with the help of crop rotation presented as a or if the circles are "drawn off" to the side and stop rotation, i.e. they will not produce soil sickness (dotted circles). For example, a fertilizer can overcome the harmful effect caused by nutrient deficiency; liming etc. can improve pH. We know such modes for two circles – "One-sided development of soil microflora" and "Toxin accumulation in soil" (dotted positions in them are absent). Allelopathy, in which soil toxins also released from the higher plants, is also a cyclic process and as it accelerates the process of soil sickness. When all effective measures are taken, allelopathy and microflora are the only factors, which causes soil sickness (Figure 1). It is interesting to note that soil sickness and allelopathy "rotate" in one direction, having common links. On the contrary, soil sickness can also reinforce allelopathy, through toxins accumulation and development of microflora.

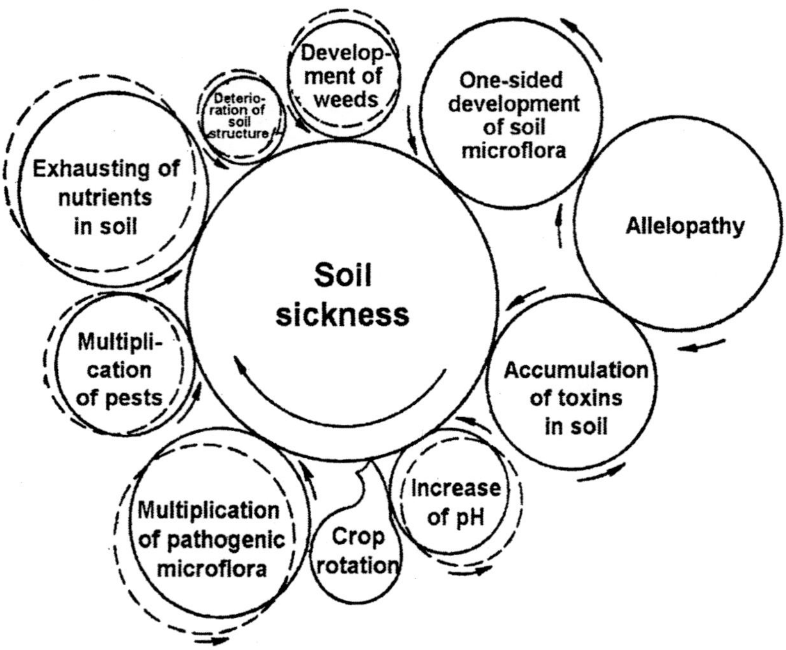

Fig. 1. Relationships between the soil sickness, allelopathy and other factors.

The toxin accumulation in soil took place not only due to excretion of root exudates of higher plants, but also as a result of soil microflora activity, which is also shown in the scheme. Thus higher plants are either suppliers of toxins or their precursors.

New terminology

In this section, some relatively new terms have been described.

I. **Kolines** (Grümmer, 1957) : These are substances of some higher plants, having chemical activity on others plants. Most frequently these are mixture of diverse, not specifically acting substances, i.e. they suppress crop growth at high concentrations and stimulate growth at low concentrations. The composition of kolines includes products of physico-chemical or biological transformation of substances initially excreted by plants, because it is practically impossible to isolate these compounds in nature.

II. **Donor/Acceptor plants : *Donor*** plants are those, which releases kolines in soil. ***Acceptor*** plants are those, which receive kolines and suffer from them.

III. **Activity:** It is the ability of plant to accumulate more or less higher concentrations of kolines, directly in its environment with the help of suitable heterotrophic microorganisms.

IV. ***Tolerance*** : It is the ability of plants to tolerate increased concentrations of kolines and even have a requirement for their presence in the environment. Allelopathic activity and tolerance are not connected, therefore following types are possible:

(i) ***Autotolerant*** *:* i.e. Transforming itself, therefore causing no soil sickness.

(ii) ***Autointolerant***: i.e. Not transforming itself, causing an acute soil sickness under its cultivation.

(iii) ***Allotolerant*** : Tolerate variable allelopathic influences and insensitive to soil sickness caused by different plant species.

(iv) ***Allointolerant*** *:* Donot tolerate other species as neighbours, prefer to be alone.

Soil sickness can have different reasons, the degree of expression, stability etc., which determines methods to overcome. Therefore, it is very important to classify the types of soil sickness. The most difficult question, in finding the reasons of soil sickness is to determine the nature of chemicals and the amounts of toxic substances in soil, due to the absence of suitable methods to extract the active substances from adsorbed state. Very little is known about the chemical nature of substances, responsible for allelopathic soil sickness.

It is not known, how far is allelopathic soil sickness spread in different types of natural and agricultural ecosystems and what is the role of microflora in the transformation of compounds in soil. Soil has an enormous, but not endless adsorbing capacity.

We should be careful not to allow increase in the quantity of toxic substances in soil; otherwise soil fertility will be irretrievably lost.

Chapter 2

SOIL SICKNESS : FIELD CROPS

N. Dzubenko

Department of Allelopathy,
Central Botanical Garden, 1, Timirvzskaya Street, Kiev, Ukraine

"It is important not to sow or to plant cereal crops on the same soils during two successive years, but to change them wisely." T. Bolotov, 1771. (Bolotov, 1952)

Since the ancient times, farmers have observed the yield reduction in plants, when they were repeatedly cultured on the same plot. That induced a more or less frequent change of cultivated fields. Monoculture is economically more profitable than entire production of each year, but it causes soil sickness. Therefore, farmers are obliged to grow many crops in rotation. Because of this elucidation of reasons of soil sickness, attracted the attention of scientists.

Long ago, Plenk (1795), de Candolle (1813) and Daubeny (1845) tried to find the reasons of decrease in soil fertility under monocultures and explained it due to accumulation of toxic substances in soil. The phenomenon of soil fertility decline was called real soil sickness. Many Experimental Stations and other Scientific Research Centres, conducted special studies on this problem. Long-term trials were established in 1842 at the Experimental Station, Rothamsted, England to investigate the cultivation of monoculture crops and are being continued till now. The trials showed that winter wheat grown in monoculture on the fertilized soil has a stable 3-3.5-fold decrease in its productivity in comparison to that grown in crop rotation also on the fertilized soil. The same was observed in relation to barley (Russel, 1936).

Monoculture: The decrease in yields of winter rye in monoculture was reported in the trials conducted from 1877 to 1923 at Halle (Germany). The yield decrease in the fifth decade, in comparison to that in the first decade constituted 51.5% on the unfertilized land and 26% on the soil fertilized with manure (Rauhe and Lehne, 1966). The experiments were established on the fields of Timiriazev Academy of Agriculture (TSHA), to study monoculture crops in 1912 in Moscow, which are being continued till now. It was found that clover and flax grown in monoculture completely stopped their development and the yield of rye and oats decreased 1.5-2.0-fold in comparison to crops grown in crop rotation, whereas the yield of potato decreased only insignificantly. Annual application of fertilizers did not remove the differences in crop yield (except potato), which has become even more significant in winter rye and oats (Yegorov, 1958, 1972).

Since 1913, on the chernozem soil of Charkov Research Station, the grain yields of winter rye grown in crop rotation without fertilizers was 9.5 dt/ha, while, with complete mineral fertilizers was 10.6 dt/ha, that of winter wheat was 5.5 and 5.5, of oats - 5.3 and 3.7 dt/ha, respectively (Narcysov, 1976).

On the fields of Mironovsk Scientific Research Institute of Wheat Selection and Seed-Production (Kiev district), the yields of sugar beet for 40-years in monoculture on unfertilized plots declined to 180 dt/ha and that of winter wheat for 30 years decreased to 10.9 dt/ha (Pyetrenko *et al.*, 1971). A great effectiveness of crop rotation was noted in the trials at the North–Western Scientific Research Institute of Agriculture (Leningrad district) (1961-1970). In rotations, the yields of winter rye, spring wheat and fibre flax increased by 34.2%, 38.1% and 124%, respectively.

A survey about the monocultures and crop rotation systems conducted by Vorobyov (1972) indicated high efficacy of crop rotation. Other investigators also mentioned a decline of yield (Zahartshenko *et al.*, 1962; Tshunderova *et al.*, 1964; Zalyalov, 1967;

Raffaelian, 1970; Burangulova *et al.* 1975; Oryel and Boyko, 1975; Dzubenko *et al.* 1977). However, the monoculture of the same crop does not always induce decrease in crop productivity. Thus, at the Agricultural Experimental Station, Poltava the cultivation of rye for 85 years, did not decline the crop productivity, though a decrease of nutrient content in soil and a decline in the activity of soil microflora were noted (Gryb *et al.*, 1973).

Dospechov (1963) reported that in long monoculture of flax on TSHA (Timiryazev Agricultural Academy) fields, under conditions of systematic fertilizing and liming had no detrimental effect on soil fertility. Plants of other species grown on flax-tired and healthy soils gave similar yields. The possibility to overcome the main biological reasons of soil sickness, using plants resistant to diseases and pests (Dospechov and co-workers, 1971). Many authors noted the stability of corn yield in long monoculture (Fedin and Biter, 1960; Tshernomaz and Korolkov, 1963; Mousyenko and Mousyenko, 1974).

Therefore, not all the plants can identically make soil sick to identical degree and can respond to the presence of toxins in it.

The problem of soil sickness continues to exist in many forms, but at present the causes of this phenomenon have not been completely explained. Dublslav (1966), when analysing the problem of soil sickness, wrote: *"The term "soil sickness" implies only something about phenomena, but nothing about their reasons, such as nematodes, fusariosis, decomposition products of roots and others".*

There are different views about soil sickness and have been grouped in to two main problems. According to first group, soil sickness is explained by deficiency of nutrients due to one-sided exhaustion of soil in monocultures (Hutchinson and Thaysen, 1918; Bronsart, 1949; Kobernus, 1951; Russel, 1955; Pryanishnikov, 1976). While according to second group, the plants suffer from the accumulation of toxic substances of plant origin in sick soils (de Candolle, 1813; Whitney and Cameron, 1903; Schreiner and Reed, 1907; Greigh-Smith, 1913, 1918; Papadakis, 1941; Patrick, 1955; Börner, 1958; Berestecki and Moroz, 1968, 1971; Grümmer, 1970; Mirchink, 1976). The plants on the exhausted soils suffer not only from the nutrients deficiency, but also from the accumulation of large amounts of toxins (Whitney and Cameron, 1903).

According to data of Ishtsheriakov (cited after Gortinsky, 1966), although the soil leached with water, loses plant nutrients but the plant yields on that soil were significantly higher. Perhaps due to leaching of water soluble toxic substances from the soil. Thus, in trials of 1908, oat plants transferred into the soil previously occupied also by oat and washed with water produced 1.5-folds higher yields than that on the natural (not washed) soil.

Periturin (1913) critically analyzing trials of Ishtsheriakov conducted a series of vegetative trials and concluded, that the experiments carried out according to the mentioned scheme of Ishtsheriakov, did not point to the presence or absence of harmful substances in soil solutions. When the soil is washed out, two mutually opposite factors - nutrition and harmful substances are removed and the result will depend on the relation between them. Additional trials conducted by the author suggested that harmful substances can appear in repeated crops. He also concluded that soil solutions also contain stimulating substances (cited after Gortinsky, 1966).

Greig-Smith (1912) pointed to insignificant accumulation of toxins in Australian soils. Their concentration depended on the type of soil, time of the year and on other reasons. Schreiner and Reed (1908) obtained a toxin from the water extracts of sick soil, the abilities of which remind wax and called it agricer. Agricer appeared to be very unstable and was quickly inactivated under the influence of sunlight. Soil disinfections by carbon disulphide or by toluol rendered agricer harmless (Grümmer, 1957). Some investigators, studying the problem of soil sickness in flax cultivation, concluded that an increase in the plant population of flax plants led to yield decrease, though the solution contained sufficient amount of nutrients for all plants (Papadakis, 1941). No yield reduction occurred, when colloids were added to the solution because the kolines excreted by living roots were adsorbed by colloids. Fastabend (1955), Koch (1955) and Grümmer (1957) considered that the causes of soil sickness of flax were excretions of toxic substances from the roots, which were accumulated in soil and inhibited the plant growth.

Börner (1958), Berestecki and Moroz (1968) and Moroz and Bukolova (1977) related the soil sickness in fruit orchards with phloridzin accumulation in roots and in soil. Krasilnikov and Garkina (1946), Berectecki (1972), Mirchink (1963) and Patrick (1955) attributed soil sickness to the presence of phytotoxic forms of microorganisms.

Russel (1955) stated that the detrimental effect of toxic root excretions including that on repeated crops to a certain degree was connected with the presence of organic substance in soil: *"The lower the content of organic substances in soil, the quicker the clover made the soil sick, which was almost definitely caused by clover roots"*.

Comparing crop rotation and monoculture, Pryanishnikov (1945) concluded why correct crop rotation was more profitable than monoculture of the same crop and gave three reasons:

1. *Chemical order* : The differences in the plant nutrition on nitrogen and ash compounds.
2. *Soil properties* : Different effects of crops on the physical properties of soil and its structure.
3. *Biological order* : Different responses of plants to pests (insects, weeds)

Bronsart (1949), Kobernuss (1951) and Mijdla (1955) believed that soil sickness was caused by reduction in micronutrients content in soil, particularly boron and molybdenum. Vogel and Weber (1931) considered that soil sickness was result of soil structure deterioration. Martin (1950b, 1960) Allen and Jensen (1951) and Kral (1924) explained soil sickness due to accumulation of nematodes and pathogenic microorganisms in soil.

From the many reasons inducing soil sickness, the accumulation of phytotoxic and inhibiting substances in soil was the initial cause determining and intensifying the influence of other factors. As it was directly related with chemical interaction of excretions, this reason is called allelopathic and the phenomenon itself – allelopathic soil sickness. We do not isolate this phenomenon from the others, but investigate it in certain interactions with other factors (microbiological, biochemical, soil nutrients status of soil, nematode accumulation etc.).

1. RESEARCH INVESTIGATIONS

Our studies were aimed at: (1) to determine the plant productivity and investigate the dynamics of allelopathic activity of soil and roots of plants growing in monoculture and in crop rotation; (2) laboratory and vegetative experiments to determine soil toxicity and the influence of yield residues of plants on the crop productivity;

(3) microbiological and nematological analysis of soil and plants; (4) investigation of enzymatic activity and nutrient regime of soil. The conduct this research, we conducted long-term field, vegetation and laboratory trials.

1.1. Study Sites

Field trials were conducted at the research stations of (a) Mironovsk Institute of Wheat Selection and Seed Science (MNIISSP), (b) Drabovsk experimental station (Cherkask district) and (c) Chartoryski experimental field (Zhytomir district). Vegetation trials and laboratory investigations were carried out at the Department of Plant Physiology, Ukrainian Academy of Science (AN USSR).

1.1.1. Mironovak Station : The fields of MNIISSP are located in the south-eastern part of Kiev district on the watershed of the rivers Ros and Rassava. The soil cover is rich in weak-humus, coarsed dust-slightly loamy and weak-leached chernozems. An agrochemical characteristic of the experimental plot is presented in Table 2, 2A. The climate is moderately continental; the average annual temperature is 7°C. The region is in the zone of moderate percipitation (Souhovyetski, 1972).

Table 2. Agrochemical characteristic of weakly-leached chernozem (by Suchowiecki, 1972)

Sampling depth (cm)	Humus (%)	Soluble nitrogen	P_2O_5	K_2O	pH_{KCl}	Hydrolytic acidity	Total of absorbed bases	Soil saturation by compounds, %
		mg per 100 g soil				mg-equivalent per 100 g soil		
0 – 20	4.18	7.9	13.5	11.4	5.8	1.71	22.1	93.6
20 – 40	3.58	6.1	12.7	10.1	6.6	1.37	26.4	95.1

Agroclimatic conditions were favourable for growing winter wheat, sugar beet, maize and other crops. Winter wheat and sugar beet were grown in monoculture at this research station for over 45 years and maize – for over 30 years. These crops were sown in the system of 10-field rotation with the following sequence: (1) perennial grasses, (2) winter wheat, (3) sugar beet, (4) maize for grains, (5) pea,

(6) winter wheat, (7) sugar beet, (8) maize for silage, (9) winter wheat, (10) barley with addition of grasses.

Table 2A. Details of field study sites

	Mirnovsk	Drabovsk	Chartoryski
District	Kiev	Cherkask	Zytomir
Soil	Weak chernozem	Weak chernozem	Forest steppe
Mean. Temp.	7°C		
Annual Rainfall (mm)		454	557
Experimental Extent period	1975-77	1970-77	1970-77
Scientist	D.J. Stepnenko	P.I. Boyko	P.I. Boyko

Mirnovsk = Institute of Wheat and Seed Science, Drabovsk = Experiment Station, Chartoryski = Experiment Station

The first four links of crop rotation and monocultures were used for investigations. The trials were carried out on unfertilized soil and soil with application of fertilizers. Sugar beet is an exception on the unfertilized soil, because mineral fertilizers during sowing are applied in rows at the following rates: N – 10, P_2O_5 – 20, K_2O – 15 kg/ha $N_{10}P_{20}K_{15}$ in monoculture as well as in crop rotation. Suitable fertilizers were also applied under this crop on the manured soil.

On the fertilized soil, manure was applied at 30 t/ha under winter wheat in crop rotation, which annually constitutes, on the average, 6 t/ha of organic fertilizers under each culture. Manure at 30 t/ha was applied under monocultural wheat only once for 5 years. On the plots of crop rotation of sugar beet and maize, the manure was applied annually at the rate of 6 t/ha.

At all research stations, the plots area was 100 m², and there were three replications. The following cultivars were sown: winter wheat – 'Mironovsk 808', 'Illichevka' (selection cultivars of B.H. Remieslo), maize – hybrid 'Bukovinskaja 3', sugar beet – one-seeded 'Jaltushko-vskaja', pea – 'Uladovsky'. O.J. Stepanenko, a scientific worker of MNIISSP, a candidate of agricultural sciences, conducted field trials during 1975-1977.

Soil samples and plants were collected 3 times during the crop season. The dates of sample collection corresponded to the following developmental phases for winter crops: (1) stem extension, (2) yellow maturity, (3) ripening; [for maize: (1) 2-4 leaf stage, (2) silking, (3) dough stage].

1.1.2. *Drabovsk Station*: Field experiments at Drabovskaja Research Station and on Chartoryski experimental field were carried out during 1970-1977 with an active participation of P.I. Boyko, a scientific worker, candidate of agricultural sciences of the Ukrainian Scientific Research Institute of Agriculture in Kiev. Dates of sample collection were the same as those in the field of Mironovsk research station. Ecological conditions of Drabovsk experimental station belong to the zone of unstable moistening of the Left-bank Forest steppe, with the average amount of precipitation 454 mm. Soils – weakly humus chernozem, coarse-silty, slightly loamy soil texture, able to float in abundant precipitation, which reduces water permeability and nitrifying soil activity. The content of humus in the arable layer was 4.2-4.5%, that of the total nitrogen – 0.22-0.25%, of mobile phosphorus – 0.14-0.18%, K_2O - 1.9-2.2%; hydrolytic acidity – 1.8-2.0 mg per 100 g of soil. During the years 1970-1973, field trials were conducted on small plots of the research station, but in 1974 they passed to large plots at the research station of the Crop Rotation Laboratory, Ukrainian Scientific Research Institute of Agriculture (UNIIZ) using the scheme below:

	Crops	Soil I	Soil II
		Unfertilised	Fertilised
1	Perennial grasses	« «	—
2	Winter wheat	« «	N_{60} P_{60} K_{60}
3	Sugar beet		20 t/ha of manure + N_{100} P_{70} K_{100}
4	Maize for grain	« «	N_{60} P_{40} K_{60}
5	Fallow (after maize for silage)	« «	—

Monoculture plots occupied by bare fallow, winter wheat and maize on the fertilized and unfertilized soils were established in 1962.

1.1.3. *Chartorysky Station:* Chartorysky experimental field is located in the north-western part of Right-bank Forest steppe in the zone of sufficient moistening. According to the data of many years, the average annual precipitation was 557 mm, 360 mm of which fall in April-September, i.e. in the period of plant vegetation. The soils are dark-grey, coarse-silty, and slightly loamy, with humus content in the arable layer of 2.7-3.3%. Samples of rhizospheric soil and plants were taken from the plots of the typical 9-field rotation with the following alternation of cultures: (1) perennial grasses – $N_{40}P_{40}$; (2) winter wheat – $N_{80}P_{60}K_{80}$; (3) sugar beet – 15 t/ha of manure, $N_{180}P_{130}K_{180}$; (4) pea – $P_{40}K_{40}$; (5) winter wheat – $N_{80}P_{60}K_{60}$; (6) sugar beet – 15 t/ha, $N_{180}P_{130}K_{180}$; (7) maize for grain – 15 t/ha of manure, $N_{120}P_{80}K_{100}$; (8) maize for grain - 15 t/ha of manure, $N_{120}P_{80}K_{100}$; (9) barley with the addition of perennial grasses – $N_{80}P_{60}K_{60}$.

The area under monocultural grain crops of winter wheat, sugar beet, maize and pea established in 1957 and is annually fertilized by the same fertilizer as in crop rotation at the rates mentioned above. However, the rates of mineral fertilizers annually applied under each crop as in crop rotation and monocultures were nearly 2-fold less until 1974 than at the present time.

Allelopathic activity of soil and plant roots was determined in newly selected samples by bioassays (Grodzinsky, 1973) at the ratio between the sample and water 2:1 and 1:10. Soil suspension was centrifuged for 8-10 min. The toxicity of soil extracts in bioassays was determined by cress seeds which are the most sensitive and radish seeds, radicals of barley and maize, as well as coleoptiles of wheat, etc.

It had been found that kolines have specific activity in different bioassays. For example, the rhizosphere of barley had the strongest inhibitory effect on barley seeds used in bioassays; soil taken from the plots under maize had its influence on maize seeds. Other crops including leguminous plants and sugar beet were very tolerant to these rhizosphere extracts. Study of the influence of soil extracts on root extension of cress, barley and maize indicates the selective ability of some test plants. Thus, the soil extracts inhibitory to cress seedlings, were found stimulatory to barley and maize seedlings.

The plants extracts contain higher concentration of water-soluble inhibitors than soil extracts, and are suitable to apply on radish seeds to determine effect on their germination in relative coumarin units (RCU) as per method of Grodzinsky (1973). Because sensitivity limits of cress seeds at high concentration of inhibitors is abruptly reduced.

1.2. Dynamics of kolines in soil and plant roots

The dynamics of kolines in the soil of Mironovsk Research Station and presented in Table 3. The length of cress roots is inversely related to the degree of allelopathic soil activity. Soil extracts had highest toxicity in 1975 – the most dry year. In the next years with more favourable climate, the inhibitory effect was significantly lower.

Soil toxicity changes at different stages of plant development. The maximum soil toxicity corresponds to the second date of collecting samples at the stage of maize flowering and yellow maturity. By the end of vegetative season, the activity of soil extract decreases. A long-term cultivation of crops of the same species on the same soil area, leads to a significant accumulation of toxins in the rhizosphere.

The maximum accumulation of inhibitors occures in crop monocultures of wheat and sugar beet on unfertilized land, especially during grain ripening. The application of organic fertilizers decreased the physiologically active substances (Kelines) in the rhizosphere of wheat and sugar beet in monocultures. It may be attributed to the improvement of physical and chemical soil properties, biochemical processes, microflora activity, changes in the quantative composition of nematodes and other factors. Soil extract from a monoculture and rotation maize crop was less inhibitory, indicating the possibility of using maize in monoculture.

A specific correlation was revealed between the activity of root and soil extracts. Thus, when sugar beet crop thinning is done, its activity in monoculture on the control and fertilized plots was equal to 49 and 39 RCU, respectively and in crop rotation was equal only to 21 and 17.2 RCU (Date I, 1975). A similar dependence has been also observed in other crops (Table 4).

Table 3. Dynamics of physiologically active substances in rhizosphere of plants from monoculture and crop rotation systems test plant : cress (% of control), Research Station, Mironovsk.

			1975			1976			1977		
Crop	Croping system	Fertilization	May 14	June 26	Aug. 28	June 17	July 27	Sept. 3	June 16	Aug. 27	Sept. 14
Winter wheat	Monoculture	Unfertilized	42.0	41.3	76.0	66.4	52.0	73.4	72.0	89.0	71.1
		Fertilized	36.2	62.0	78.4	80.8	54.2	69.2	84.0	89.5	83.0
	Crop rotation	Unfertilized	100.0	98.3	86.6	92.0	67.4	83.0	92.0	94.3	85.0
		Fertilized	91.8	93.9	84.2	86.7	57.9	77.4	124.0	94.5	93.5
Sugar beet	Monoculture	Unfertilized	68.9	74.5	87.1	70.2	43.4	77.0	84.0	78.4	72.3
		Fertilized	61.5	73.2	85.6	71.2	63.5	87.0	92.0	84.0	76.4
	Crop rotation	Unfertilized	68.9	80.6	83.2	75.5	68.1	87.9	94.5	85.7	80.2
		Fertilized	81.5	84.7	91.8	73.9	65.8	82.2	101.0	88.0	87.0
Maize	Monoculture	Unfertilized	89.2	93.0	86.1	75.5	71.3	82.2	88.0	101.0	91.0
		Fertilized	71.6	101.0	80.4	71.2	65.8	79.8	99.0	99.4	73.2
	Crop rotation	Unfertilized	64.8	97.6	83.2	85.1	78.2	90.7	97.0	106.7	79.0
		Fertilized	95.2	104.4	82.3	73.4	62.7	—	88.0	95.5	86.4

Table 4. Dynamics of water-soluble inhibitors in plant roots from monoculture and crop rotation systems, Research Station, Mironovsk, 1976 (Relative Coumarin Units)

Crop	Crop system	Fertilization	Date of sampling		
			April 17	July 27	Sept. 3
Spring wheat	Monoculture	Unfertilized	19.4	33.0	14.0
	Monoculture	Fertilized	27.0	29.0	17.8
Winter wheat	Crop rotation	Unfertilized	10.7	23.0	11.3
	Crop rotation	Fertilized	24.0	28.0	13.5
Sugar beet	Monoculture	Unfertilized	115.0	69.0	33.0
	Monoculture	Fertilized	85.0	74.0	49.0
	Crop rotation	Unfertilized	67.0	54.0	27.5
	Crop rotation	Fertilized	65.0	52.0	29.7
Maize	Monoculture	Unfertilized	110.0	45.0	27.2
	Monoculture	Fertilized	95.0	36.0	28.2
	Crop rotation	Unfertilized	80.0	41.0	23.0
	Crop rotation	Fertilized	92.0	32.0	—

When studying the dynamics of allelopathic activity of root extracts, these were found dependent on their developmental stages and on the conditions of plant growth. Before the beginning of the reproductive development stage, the content of physiologically active substances increases in plants.

In dry year 1975, when rains in May, July and August was only 50% of the average rains, the amount of kolines in plant roots was higher than that in the normal rainfall years 1976-1977 (the kolines amounts in the summer months exceeded the average long-term rates by 25-30%).

Plants grown in monoculture had higher contents of inhibitors than rotation crops. However, not all the plants with many physiologically active substances (e.g. maize) were characterized by high allelopathic activity and make the soil sick. A high activity of inhibitors in the root extracts of maize did not increase soil toxicity and did not cause sickness in long-term maize cultivation in monoculture.

Observations on the dynamics of the increase of aboveground plant matter (Table 5) displayed reverse dependence of the plant growth on the soil allelopathic activity. However, not all the plants similarly respond to a monoculture. The largest decrease in biomass was noted in winter wheat and sugar beet monocultures.

Table 5. Dynamics of aboveground biomass accumulation in monoculture and crop rotation systems, Research Station, Mironovsk, 1975

Crop	Crop system	Fertilization	Date of sampling					
			May 14		June 26		August 28	
			g	%	g	%	g	%
Winter wheat	Monoculture	Unfertilized	1.7	34	1.4	47	-	-
		Fertilized	5.3	106	2.4	80	-	-
	Crop rotation	Unfertilized	5.0	100	3.0	100	-	-
		Fertilized	5.2	104	3.2	107	-	-
Sugar beet	Monoculture	Unfertilized	0.2	22	31.6	50	115	60
		Fertilized	0.7	77	114.6	181	136	70
	Crop rotation	Unfertilized	1.1	100	63.3	100	193	100
		Fertilized	1.4	130	191.2	302	256	133
Maize	Monoculture	Unfertilized	0.5	84	108.2	79	520	80
		Fertilized	0.9	160	289.6	211	812	125
	Crop rotation	Unfertilized	0.6	100	137.3	100	650	100
		Fertilized	0.5	98	170.6	124	780	120

g : Biomass g per plant,
% : per cent over crop rotation without fertilization

The importance of plant cultivation system was shown dry year 1975. In monoculture plots, the crop plants were weak. Plots in crop rotation were quite satisfactory and were not resown. This was not owing to insufficient soil moisture, because the productive moisture content in crop rotation plots in the 0-10 cm layer was 4.3 mm as against 8.3 and 17.3 mm in monoculture plants. Therefore, the reason of bad overwintering of monocultural winter crops was attributed to allelopathic factor. The yield data (Figs. 2-4) also indicate significant decrease in productivity of monocultural crops.

The application of manure, slightly reduced the detrimental effect of monoculture on plant productivity of sugar beet and winter wheat, but did not eliminate it completely. At the same time, the growth and yield of maize monoculture on the fertilized soil was slightly better in 1975. So, not all the plants decrease their productivity to the same degree under conditions of monoculture (excluding deficiency of nutrients in soil).

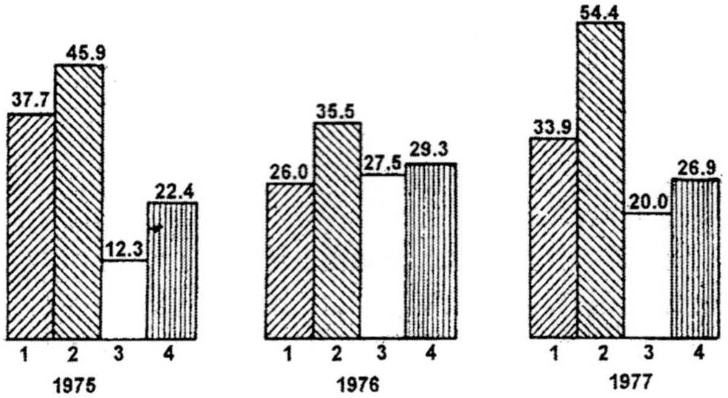

Fig. 2. Diagram of fertility of winter wheat (Mironovsk station, dt/ha). 1 – crop rotation without fertilization, 2 – crop rotation with fertilization, 3 – monoculture without fertilization, 4 – monoculture with fertilization.

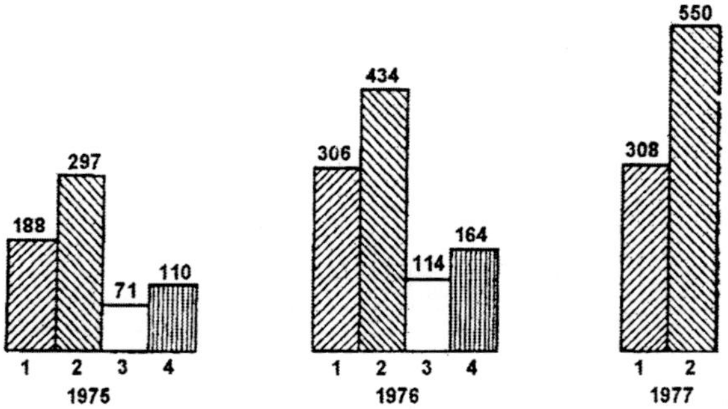

Fig. 3. Diagram of fertility of sugar beet (Mironovsk station, dt/ha). Legends as in figure 2.

The yield stability of maize under conditions of long-term monoculture (Tshernomaz and Korolkov, 1963; Mousyenko and Mousyenko, 1974) have been also confirmed in our experiments. An insignificant phytotoxicity of maize was also indicated by allelopathic activity of water extracts. Only applying organic fertilizers cannot eliminate the phenomenon of soil sickness. We have found very high kolines content in monoculture crops also in Chartoryski experimental field. Soil toxicity was maximum at the yellowing and full maturity stages of winter wheat and pea and at the maize flowering, which corresponds to the second terms of sample collection. In spring and autumn, the content of kolines in soil as well as in plants decreased abruptly. On this basis, it can be suggested that the phenomenon of soil sickness was closely related to the formation of various physiologically active compounds and their excretion through roots into environment, which influence the mineral nutrition and total metabolism of plant cenosis directly as well as through soil microorganisms (Ivanov 1973).

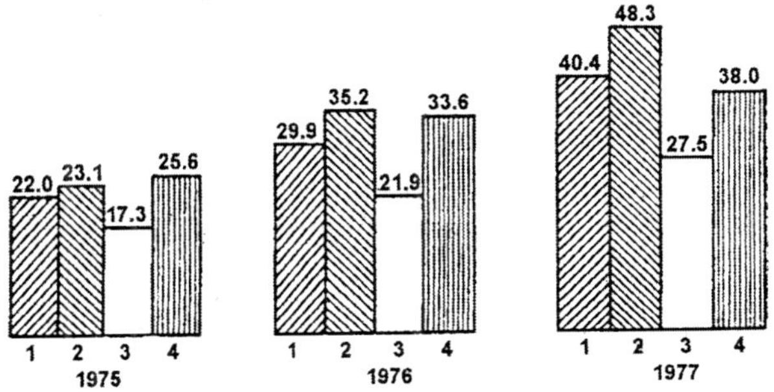

Fig. 4. Diagram of fertility of maize (Mironovsk station, dt/ha). Legends as in figure 2 and 3.

The study of the dynamics of inhibitors in plant roots grown in monoculture and in rotation crops indicates a change in their content depending on the environment. Arid conditions of 1975 increased the activity of kolines in plant roots, particularly in sugar beet varieties (Table 6).

Table 6. Dynamics of physiologically active substances in roots of plants from monoculture and crop rotation systems (Relative Coumarin Units), Chartoryski research field

Crop	Crop system	1973			1975		
		June 13	July 17	Oct. 9	May 28	June 23	Sept. 24
Winter wheat	Monoculture	92	21.6	-	33.5	56.2	-
Sugar beet	Crop rotation after peas	60	9.5	-	17.5	30.8	-
	Monoculture	137	263.0	88.0	39.7	187.0	147.0
Maize	Crop rotation after wheat	160	141.0	80.0	20.7	115.0	91.0
	Monoculture	85	19.3	38.0	112.0	44.0	16.4
	Crop rotation after maize	135	34.0	22.5	110.0	43.0	12.6
	Crop rotation after sugar beet	137	20.2	24.0	98.0	70.0	30.0
Peas	Monoculture	190	23.5	-	85.0	117.0	-
	Crop rotation after sugar beet	55	17.5	-	32.8	56.0	-

The highest activity of root extracts was during the grand growth phase, i.e. before the reproductive phase in plants. At the full ripeness stage, the content of inhibitors in roots decreases drastically. Plants growing in monoculture accumulated significantly more inhibitors than plants of rotation crops. A high content of inhibitors was found in the roots of sugar beet, winter wheat and pea grown in monoculture, in comparison to the growth of these crops in crop rotation.

Some deviation was observed only in maize plants during the periods of sample collection I and II. This was explained by very high level of kolines in the roots of preceding crops at the end of the vegetative season. In autumn, they were washed out (leached) by rains and in spring they again come into the plants. Therefore, maize crop caused some degree of soil sickness.

The crop productivity was correlated with soil toxicity. Results of long-term experiments indicated that the productivity of crops grown in monoculture was decreased in comparison to rotation crops.

The strongest growth inhibition in above ground plant parts and the yield were noted on the plots under winter wheat, sugar beet and pea grown in monoculture (Table 7). The yield loss was significantly lower in the plots of maize grown in monoculture than in pea and winter wheat. The highest increase in yield was on the plots, where pea was grown in crop rotation. There was great difference in the content of inhibitors in the roots of pea grown in monoculture and in crop rotation.

In the treatments with maize, where no large `increase in' yield was found than monoculture, there were no special differences in the koline content between monoculture and crop rotation. The dynamics of water-soluble inhibitors in the rhizosphere of monoculture and rotation crops growing at Research Station, Drabovsk showed some deviations from the fields data in the fields of Mironovsk Scientific-Research Institute. Sometimes we did not find clear stable differences between to koline content in soil extracts taken from crop rotation and monoculture. This is possibly explained by a relatively young age of monoculture crops (since 1964), which had no time to accumulate the sufficient quantity of toxins in soil.

Table 7. Productivity of crops in monoculture and crop rotation systems, Research Field Station, Chartoryski

Crop	Cropping system	1973		1974		1975		1976		1977	
		yield (t/ha)	%	yield (t/ha)	%	yield (t/ha)	%	yield (t/ha)	%	yield (t/ha)	%
Winter wheat	Monoculture	2.18	100	2.25	100	1.28	100	2.57	100	1.67	100
	Crop rotation	3.66	169	4.01	178	3.33	260	5.48	213	4.75	284
Sugar beet	Monoculture	15.50	100	28.70	100	28.00	100	34.20	100	34.9	100
	Crop rotation	21.60	139	36.80	128	48.20	172	37.00	108	42.5	120
Maize	Monoculture	3.94	100	2.25	100	3.21	100	1.70	100	1.56	100
	Maize - maize	4.60	117	3.29	146	4.58	143	2.32	136	2.63	170
	Sugar beet - maize	4.88	124	3.53	157	5.79	180	2.56	150	1.85	120
Peas	Monoculture	1.10	100	0.16	100	9.1	100	1.12	100	0.39	100
	Crop rotation	2.58	232	1.12	700	24.5	170	3.82	341	3.29	845

Long-term studies conducted to explain inhibitory activity in plant roots have shown that water extracts of sugar beet and winter wheat grown in monoculture were most active allelopathically. Koline activity in the roots of winter wheat in monoculture, at date I of sampling constituted 68 RCU in monoculture and 37 RCU in crop rotation. This indicated the possibility of plants responding to sick soil. The application of organic fertilizers (manure at 15 t/ha) and complete mineral fertilizer, decreased the kolines content in the soil and roots of winter wheat monoculture. Indices of plant production were correlated with soil toxicity. Thus, the average weight of winter wheat plants grown in monoculture on unfertilized land was 4.5 g on June 17, 1975 as compared to 5.5 g, with the application of the complete mineral fertilizer. In crop rotation it was 9.3 g and on the fertilized soil - 11.6 g, respectively. The poor germination of winter wheat on continuous plots were additionally sown with barley in 1976. The reasons of that were the same as those at Mironovsk research station. Crop productivity of monocultures was also 40-50% lower than that in crop rotation. The application of manure increased the productivity of monocultures. Thus the application of organic fertilizers i.e., change in nutrition conditions and soil microflora can contribute to a certain degree for the removal of soil toxicity.

The activity of water extracts from the soil does not always correlate with its sickness, hence, we used the method of direct bioassay of Neubauer-Schneider (Grodzinsky and Grodzinsky, 1973).

Laboratory trials were conducted with the soil samples, collected from monocultures and crop rotation. These samples were also used for determination of water-soluble inhibitors and microbiological and biochemical analyses. The soil was sieved through a 2 mm net sieve and rootlets were carefully separated. Each Petri dish was filled with 50 g sand and then was 80 g soil added. The rye cv. 'Podolansky', barley cv. 'Nosovsky 2', wheat cv. 'Mironovskaya 808' were used as acceptor plants. Seeds were pre-germinated in thermostat and then 12 seeds were placed Petri dish. Before planting soil was moistened with 30 ml of distilled water. The trials were conducted in 3 replications. Biometric measurements of plants grown on the soils taken from under monocultures and crop rotation were regularly made and at the end of the experiments the plant weight was taken into consideration. The trials lasted for 20-25 days.

Table 8. Effect of soil toxicity on different test crops in bioassays at Research Station, Drabovsk, 1972

Crop	Cropping system	Rye								Barley			
		July 27				September 16				September 16			
		Fresh weight		Dry weight		Fresh weight		Dry weight		Fresh weight		Dry weight	
		g	%	g	%	g	%	g	%	g	%	g	%
Winter wheat	Monoculture	0.34	82.0	0.06	96.7	0.52	93.0	0.07	88.0	0.39	77.3	0.05	89.2
	Crop rotation	0.41	100.0	0.09	100.0	0.55	100.0	0.08	100.0	0.52	100.0	0.08	100.0
Sugar beet	Monoculture	0.37	68.0	0.05	70.0	0.64	95.0	0.07	98.0	0.54	94.3	0.07	98.6
	Crop rotation	0.55	100.0	0.07	100.0	0.68	100.0	0.09	100.0	0.58	100.0	0.07	100.0
Maize	Monoculture	0.36	72.8	0.05	75.7	0.35	81.0	0.06	90.0	0.34	79.0	0.05	93.0
	Crop rotation	0.49	100.0	0.07	100.0	0.41	100.0	0.07	100.0	0.43	100.0	0.06	100.0
Black fallow	Monoculture	0.41	86.3	0.07	95.0	0.42	70.0	0.07	104.0	0.69	94.6	0.10	82.0
	Crop rotation	0.46	100.0	0.06	100.0	0.61	100.0	0.07	100.0	0.74	100.0	0.12	100.0

g = Weight g per plant, % = per cent over crop rotation.

The vigour of plants in direct bioassay indicated the presence of inhibitors in soil. Thus, in the soil taken from the fallow (crop rotation) plots, the increase in above ground parts biomass of acceptor plants was quite high. This was especially important, when barley is used as a test object (Table 8). However, monoculture soil exerted inhibitory effect on the plant growth indices, which was explained by suppression of microflora and by the changes in the biochemical processes in soil. Due to absence of plant residues on fallow plots, the biological soil activity decreased. This was indicated from data of quantitative composition of microflora and the level of enzymatic activity of soil. Therefore, when conditions were favourable, one season fallow field in crop rotation was enough for decomposition of plant residues and soil intoxication.

The extent of inhibition of test-plant growth was clearly visible in the first 5-10 days of the trials. Subsequently, the differences between the treatments began to level off, because the plant was most sensitive during seed germination and the initial stage of development. The growth inhibition was displayed not only in the plant height and shoot biomass, but also in the root system. Barley plants displayed higher sensitivity. The increase in the shoot biomass in the soil under monoculture was significantly lower than that in crop rotation, because the soil from monoculture indicated a higher toxicity than that from crop rotation.

Indirect bioassay studies, soil toxicity was correlated with the indices of koline activity in aqueous soil extracts, except with monocultural pea (Chartorysky experimental field). The water soil extracts from pea monoculture, displayed very high allelopathic activity to cress and radish seeds, but when grown directly in the soil of rye and barley, the pea stimulated their growth (the average dry weight of plants grown on the soil under pea monoculture was 3.13 g and that from the soil under crop rotation was 2.33 g). It was also bene for longer duration of plant growth under conditions of a vegetation experiment in soil culture, where maize was a test crop. Similar results were obtained when plants were growing in water culture (Dzubenko, 1966).

Comparing the results of laboratory experiments with the data on the dynamics of growth of above ground plant parts (Table 5), we found stronger effect of plant growth inhibition, when plants were grown directly in the field. This was probably explained by a partial inactivation of inhibitors, which lost their activity after some time, and also by a substitution of test-objects interrupting conditions of monocultural growth of one and the same crop culture. The use of wheat as a test-object indicated its higher sensitivity, when it was grown on soil taken from the rhizosphere in comparison with soil from under sugar beet and maize.

For example, the weight of dried maize plants grown on the soil of maize monoculture (Chartorysky experimental field, 1973), was 10.8 g, whereas, in soil from pea crop rotation it was 25.2 g. Such regularity also repeated in other treatments, where soil was taken from maize crop. Despite the fact that maize in field trials did not induce strong soil sickness under conditions of vegetative experiments, where koline concentration increased due to limitations of soil volume, the effect of autotoxicity was extremely significant. There are no basis to explain this phenomenon due to nutrients deficiency in the soil, because the content of nitrogen, phosphorous and potassium in soil was quite sufficient.

1.3. Nematodes

Considering significant role of nematodes in the life of biogeocenosis, some scientists have correlated their population with the phenomenon of soil sickness (Martin, 1950, 1960; Allen and Jensen, 1951; Kral and Mijdla, 1965). Sigareva began the nematological investigations in 1973 and proceeded from the possibility to use the number of phytonematodes of different ecological groups as a test for soil toxicity. The composition of nematodes groups was mainly presented by saprobionts, indicating intensification of decomposition processes by mycohelminthes (indicators of fungus decomposition) and also by phytohelminthes (initial cause of plant diseases – phytohelmintoses). Thus phytohelminthes were important in plant life. The accumulation of nematodes in the roots and root zone were harmful to plants and decreased their yield and can cause soil sickness.

Study of phytonematodes population in soil and in plant roots grown under conditions of vegetative trials on Chartorysky field were conducted in October 1973. The composition of phytonematodes fauna in crop rotation in the adjacent fields was found same independently of the crop species grown. However, the population of individual species was determined by the composition of plants, their physiological condition and the toxicity of their roots and soil in different vegetation seasons.

Study of the dynamics of species-specific and numerical composition of nematodes on the roots of maize grown under conditions of a vegetative experiment showed that after harvesting the main crops, the soil was saturated by residual nematode populations, whose density depends upon the crop species grown. Out of 44 nematode species found in Chartorysky field, only 15 species were found in the vegetative trial. Three nematode species – *Acrobeloides buitschilii, Aphelenchus avenae* and *Pratylenchus pratensis,* respectively from saprobionts, mycohelminthes and phytohelminthes ecological groups were discovered in all samples. The population of other species was similar. The mycohelminthes group included the largest number of species. The populations of mycohelminth *Aphelenchus avenae* was significant in all treatments, but their maxiumum density in the soil was in plots of sugar beet (430-609 individuals/g of roots) and maize (123-220 individuals/g). The transfer of nematode infested soil to favourable temperature conditions and to maize crop led to a rapid outbreak of mycoses in its roots and to subsequent multiplication of *A. avenae* feeding on the fungi. The highest density of saprobionts was observed in the pots with soil from maize and sugar beet plots, where the mycohelminthes were dominant. The roots infected by mycoses decay, which results in the development of conditions for multiplication of nematodes-saprobionts.

The third, most important group of nematodes-phytohelminthes present in our experiments was *Pratylenchus pratensis*. Its high population were observed nearly in all treatments, but the maxiumum density was in the pots with soil from the virgin plot with wheat and maize monoculture (36.7 and 47.6 individuals/g, respectively). The highest population of *P. pratensis* was left in the soil by wheat and other cereal grasses in the virgin soil, indicate that these plants during the vegetative period create favourable conditions for its multiplication, i.e. they were its accumulators. While complete absence of this species was found in soil after harvest of sugar beet in monoculture.

The initial causes of the development of *A. buitschlii* population in pots were mycoses, which decomposed the roots. Because this phenomenon was secondary and short lived, the population of saprophyte remains lower then mycohelminthes. Therefore, the reverse correlation between soil toxicity and the population of nematodes was correct only for *P. pratensis,* because only this species was directly connected to root system of host-plant and its physiology.

1.3.1. *Mironovsk Station* : The study of the number of phytonematodes in soil and in plant roots grown in a monoculture at the research station Mironovsk was conducted in the summer-autumn months of 1976 as well as in the spring-summer months of 1977. The analysis of the obtained data showed that Certain species of parasitic nematodes were accumulated in crops grown in monoculture also at Mirnovsk station.

I. Sugarbeet

The sugar beet in monoculture was infected by beet nematode and the *Pratylenchus* disappears completely. However, the populations of *Pratylenchus* (to 500 individuals per 100 g of soil), stem nematodes (100) and *Helicotylenchus* (270 individuals) increased. Groups of mycohelminthes and saprobionts were not numerous during the July and September, which indicated the absence of decomposing processes in roots (Table 9, 10).

The dominance of phytohelminthes in the roots and rhizosphere of sugar beet monoculture indicated that these nematodes contributed to the soil sickness. In 1977, the main cause of poor condition of beet in monoculture was the infestation with beet nematode. A significant number of viable cysts were found in unfertilized plots. A high density of parasitic nematodes in the soil of crop rotation plots (300 individuals per 100 g of soil) caused failure of sugarbeet after sprouting and increased the infestation with rootworm. The preceding winter wheat crop left large population of *Pratylenchus, Helicotylenches* and *Tilenchorinchyses* in the early spring in unfertilized plots for the preceding sugar beet. During summer period, the population of *Pratilenchus* were reduced in sugar beet plots in crop rotation system, but the population of beet nematodes increased in monoculture. In June, the number of phytohelminthes increased in the rhizosphere of sugar beet monoculture to 225 individuals in unfertilized land, 105 in fertilized land and 30 and 50 individuals per 100 g of soil in crop rotation, respectively.

Table 9. Nematode population in plant roots in monoculture and crop rotation systems, Research Station, Mironovsk (individuals/g of roots, 1976) (by D.D. Sigareva)

Nematode spp. and ecological group	Wheat				Maize			Sugar beet			
	July 28				September 2			September 2			
	Monoculture		Crop rotation		Monoculture		Crop rotation	Monoculture		Crop rotation	
	Unfer-tilized	Ferti-lized	Unfer-tilized	Ferti-lized	Unfer-tilized	Ferti-lized	Unfer-tilized	Unfer-tilized	Ferti-lized	Unfer-tilized	Ferti-lized
Ditylenchus dipsau	9.5	2.5	3.2	23.8	27.6	100.0	5.7	-	-	-	-
Prachylenchus nanus	-	-	-	-	-	-	-	-	1.4	-	-
P. pratensis	-	-	1.6	2.8	0.8	5.0	-	-	-	-	-
Heterodera schachtii	-	-	0.8	-	-	-	-	1.0	0.8	-	-
Helicotylenchus dehystera	-	1.0	-	7.0	0.4	-	-	-	-	-	-
Tylenchorhychus dubius	-	0.5	-	-	-	-	-	-	-	-	-
Longidorus	-	-	-	-	-	-	-	-	-	-	-
Phytohelminthes	9.5	4.0	5.6	33.6	28.8	6.0	5.7	1.0	2.2	-	-
Saprobionts	58.0	17.1	22.4	44.8	9.4	1.0	2.7	4.0	2.6	13.2	7.7
Mycohelminthes	37.0	23.5	11.6	19.6	13.6	4.0	11.1	4.0	1.4	6.6	2.2
Total of nematodes	**104.0**	**44.6**	**39.6**	**98.0**	**51.8**	**11.0**	**19.5**	**9.0**	**6.2**	**19.2**	**9.9**

Table 10. Nematode number in plant roots in monoculture and crop rotation systems, Research Station, Mironovsk (individuals /100g of soil), July 28, 1976 (by D.D. Sigareva)

Nematode spp. and ecological group	Wheat				Maize				Sugar beet			
	Monoculture		Crop rotation		Monoculture		Crop rotation		Monoculture		Crop rotation	
	Unfer-tilized	Ferti-lized	Unfer-tilized	Ferti-lized	Unfer-tilized	Ferti-lized	Unfer-tilized	Ferti-lized	Unfer-tilized	Ferti-lized	Unfer-tilized	Ferti-lized
Ditylenchus dipsau	12	30	6	200	24	42	12	70	-	-	-	40
Prachylenchus nanus	-	10	54	40	-	-	-	100	45	60	190	-
P. pratensis	60	10	120	100	12	54	-	-	-	-	-	10
Heterodera schachtii	-	-	-	-	-	-	5	-	565	828	-	-
Helicotylenchus dehystera	70	30	110	60	60	130	15	140	-	-	100	110
Tylenchorhychus dubius	350	740	110	380	-	12	2	90	-	-	5	140
Longidorus	10	-	-	-	-	-	-	-	-	-	-	-
Phytohelminthes	520	820	400	780	92	238	34	400	610	868	340	300
Saprobionts	384	425	214	320	30	102	40	230	135	580	140	250
Mycohelminthes	360	370	200	480	60	330	56	380	115	230	200	130
Total of nematodes	**1246**	**1615**	**814**	**1580**	**182**	**670**	**130**	**1010**	**860**	**1680**	**680**	**680**

II. Wheat

Wheat was also infested by several species of parasitic nematodes. When winter wheat was grown in crop rotation on fertilized land, the maximum were the stem nematodes (23.8 individuals/g of roots in July which increased to 185 individuals in September) in the same plot. Hence, it was confirmed that root residues in soil were the main sources of endoparasitic nematodes. Besides large population of *Tylenchorhynchus* and *Helicotylenchus* (about 740 individuals per 100 g of soil) were present in the rhizosphere of wheat. The first one was dominant in monoculture, whereas the second one in crop rotation (Table 10). The analysis of number of nematode in the roots and soil of winter wheat during harvest period shows that they were dominant in monocultures. A significant increase of mycohelminthes, also caused an increase in fungus population in these plots.

In the summer season, the *Pratylenchus pratensis* population was dominant in wheat roots (33 individuals/g of roots). The phytohelminthes were also found in monoculture of wheat (29.5-43 individuals/g of roots and about 105 individuals/100 g of soil).

Crop rotation plot of winter wheat without fertilizers was most infested and the number of phytohelminthes in soil increased up to 444 individuals per 100 g. In this period the population of root nematodes decreased, because root system of wheat almost disappeared and nematodes move into the soil.

Therefore, winter wheat favoured the gathering of stem nematodes and *Helicotylenchus* i.e. species which infect the next crop i.e. sugarbeet in rotation. Large population of endoparasites of *Pratylenchus* and exoparasitic nematodes of genera *Helicotylenchus* and *Tilenchorinchus* in the wheat plot (especially unfertilized) may be the cause of root rot and emptiness of winter wheat spikes.

III. Maize

The maize crop soil, appeared to be the least infested by nematodes. The stem nematodes (27.6 individuals/g of roots) was studied in unfertilized monoculture plots. *Pratylenchus* (15 individuals/g of roots) were comparatively dominant in the fertilized plots of monocultures. Both the species are exoparasites and mutually exclude each other. If wheat was infested with two parasites :

Tylenchorhynchus and *Helicotylenchus*, then only the last one occurred in maize (mainly in monoculture).

Roots of maize monoculture increased the total nematode number, with predominance phytohelminthes and mycohelminthes (Table 9). At the reproductive phase, the number of phytohelminthes in the soil of unfertilized or fertilized monoculture plots was higher in comparison to fertilized or unfertilized crop rotation system.

Maize at one research station was less infested with phytohelminthes, but this crop contributed to preservation of high population of *Pratylenchus* and *Helicotylenchus*. Its long vegetative period in the fields significantly favour the build-up of nematodes.

1.3.2. *Chartorysky Station* : Their were 44 species of phytonematode in soil and plants roots in Chartorysky field during May-September 1973. Their highest population was found in wheat monoculture (71 individuals/g of roots), fodder grasses (25-68) and pea (117).

The nematode number was considerably lower in maize (8 individuals/g of roots) and sugar beet (8.8). The crops decreased the number of all nematodes particularly phytohelminthes. Thus, the number of *Pratylenchus pratensis* in wheat was 55 individuals /g roots, in clover – 14, in maize it did not exceed 3.4, whereas, in sugar beet it was still lower – 2.7 individuals/g root. A more detailed description had been given by Dzubenko and co-workers (1973). The stem nematode are found quite evenly and in insignificant number in all the plots, except barley and achieved a higher population in wheat monoculture (8.8 individuals/g of soil).

Maize was least infested with phytohelminthes. Only in crop rotation at the beginning of vegetation (May, June) sufficient numer of *P. pratensis* (2.8-3.4 individuals/g) was present in its roots and then like other species of phytohelminthes it almost disappeared. Sugar beet in monoculture, despite a weak infestation with *P. pratensis* and *D. dipsaci* was parasitized by *P. nanus*. The distribution of phytoparasitic nematodes in the first half of vegetation season (May, June) depended on the amount of rainfall. While in the second half of vegetative period, the physiological condition of the studied crops influenced the number of phytohelminthes. Even the sufficient soil moisture in July could not increase the population the nematodes in ageing plants. The increase in plant toxicity was also connected with the increase of parasitic nematodes e.g. wheat. The highest amount of physiologically active substances in June in the wheat grown in crop rotation after pea and in monoculture (285-323 RCU) coincided with the maximum root inhibition in that period by *Pratylenchus* (40.6 and 55 individuals/g) and vice versa. While in the plot of winter wheat grown after clover, whose roots were free of *Pratilenchus*, the toxicity was significantly lower (105 RCU). In wheat monoculture plot, the population of saprobionts and mycohelminthes were maximum at the beginning of vegetation (May) - 163 and 16.7 individuals/g, respectively, because large amount of roots died during the winters. Root toxicity was lowest in saprobiont and was maximum in micohelminthes.

1.3.3. *Drabovsk Station* : Investigations of the number of phytonematodes in the soil and roots of plants grown in crop rotation system and in monoculture at Drabovsk experimental station were carried out in May – September of 1973. The investigation showed that the nematode population was less in soil than in the plant roots. Besides, nematodes in soil samples were distributed evenly and did not exceed (0.2-8.5 individuals/g) in any plot. In monoculture plots at the end of vegetative growth, the population density of nematodes of all ecological groups increases slightly.

I. Wheat

The influence of species specificity of plants on the composition of nematodocenoses was clearly evident in the roots. Thus, significant populations of saprobionts (73.2 individuals/g) and mycohelminthes (19.2 individuals/g) were buildup in the roots of winter wheat. Phytohelmintes were mainly represented by one species *Pratylenchus penetrans* and achieve maximum population (41 individuals/g) in July. Their population did not decrease at the end of vegetative phase.

In wheat monoculture, the maximum population of saprobionts (163.9 individuals/g) and micohelminthes (16.7 individuals/g) was observed at the beginning of vegetative phase, which was probably connected with the decomposition of dead roots during the winter season. Then their number decreased in June (17.6 and 2.2 individuals/g) and again increased by the end of vegetative phase (22.8 and 71.1 individuals/g). While in plot of crop rotation, the maximum population of *Pratylenchus penetrans* was 55 individuals/g in June.

Comparing these densities to experimentally obtained data with "a permissible limit" (the amount of nematodes per 1 g of soil, at which a noticeable yield decrease occurs), for *Pratylenchus penetrans* it was 100 individuals per 300 mg of soil (Hoestra and Oostenbrink, 1962), for *Pratylechus* spp. – 100 individuals per 200 mg of soil (Brown, 1964). Hence the population of *Pratylenchus penetrans* observed by us in the wheat roots significantly decreased the yield. Of all the examined cultures, the high population of *Pratylenchus penetrans* was observed only in the wheat roots. Nematodes from saprobiont and mycohelminthes groups also had higher population.

Roots of sugar beet were entirely free of nematodes. The small nematode population of all ecological groups (saprobionts – 6.5 individuals, mycohelmintes – 4, phytohelmintes – 2.5 individuals), at the beginning, completely disappeared at the end of vegetation. The nematode fauna of maize was also very poor. Of considerable interest was the nematode fauna of pea, during the vegetative phase, the number of saprobionts and phytohelmintes in the crop rotation plots and in monoculture increased to 121.5 and 45.9 individuals, respectively.

The nematode population of roots significantly exceeds the soil population, it indicates that nematodes were attracted to plant roots. The number of phytonematodes in roots defends on the host-plant species. For example, the maximum population were observed in the roots of wheat and barley, smaller population – in the roots of clover and pea, while, roots of sugar beet and maize were almost nematode-free.

The results of nematological investigation in crops from Chartorysky experimental field and Drabovsk experimental station (MNIISSP), showed that maximum multiplication of phytohelmintes and mycohelminthes was found in winter wheat. The monocultures of sugar beet in the fields of Research Station, Mironovsk were infected by beet nematodes. At Experimental Station, Drabovsk, the maize roots and rhizosphere were nearly nematode free. The phytohelminthes were dominat in monocultures, which decreased the yields of monoculture crops.

1.4. Soil enzymes

When finding the causes of the soil sickness phenomenon, we find the disturbance in the relationship between the individual plant nutrients due to one-sided soil exhaustion in monoculture. However, the increase in fertilizer rates in the long term did not provide high yields permanently, and the yields coninued to decrease. This means that soil fertility cannot be substituted by the application of fertilizers. Hence the main reason of depressive effect of monoculture systems consisted of fertility decrease due to biochemical processes in soil, which had not been investigated sufficiently. The biochemical processes in soil depended to a significant degree on the level of enzymatic activity. Practically all biochemical processes in soil were carried out with the participation of enzymes. Many investigators (Galstyan, 1963; Haziyev, 1972, etc.) consider enzymatic activity as an indicator of biological soil activity and, therefore, of fertility as well. However, the activity of any enzyme cannot fully characterize the total enzymatic activity of soil and it is impossible to conclude about soil fertility on the basis of any enzyme (Kouprevitsh, 1958). Some authors try to explain the relation between soil enzymatic activity and its fertility on the basis of correlation between enzymatic activity and the number of microorganisms.

The aim of our studies was also to explain the participation of enzymes in the biochemical processes in relation to the microflora activity and varying level of soil fertility. We have determined the activity of dehydrogenase, polyphenoloxidase, catalase and peroxidase from the group of redox enzymes. These enzymes actively participated in the decomposition and synthesis of organic soil substances and influenced two processes (a) mineralization of plant residues and (b) formation of humic substances. The dehydrogenases play significant role in the decay and synthesis of organic substances.

1.4.1. ***Dehydrogenases*** **:** Dehydrogenases catalyse the reactions of dehydrogenation, i.e. the removal of hydrogen from organic compounds and fulfill the role of intermediate hydrogen carriers. Different carbohydrates, organic acids, alcohols, humic acids, etc. can be substrates for dehydrogenation. The quantity of dehydrogenase activity was determined by the method of Galstyan (1952). Dehydrogenases were most active when manure was applied (Table 11). As a rule, the activity of this enzyme increased with change in soil pH towards alkaline reaction.

Table 11: Dynamics of dehydrogenase activity (mg formazon 100 g^{-1} soil day^{-1}) in rhizosphere of crop plants, Research Station, Mironovsk.

Crop	Cropping system	Fertilization	1975			1976			1977		
			May 14	June 26	Aug. 30	May 17	July 27	Sept. 3	June 16	July 27	Sept.14
Winter wheat	Monoculture	Unfertilized	30.0	11.5	32.8	14.0	8.7	1.5	12.0	13.5	8.7
	Monoculture	Fertilized	29.0	37.8	25.0	10.0	10.3	9.3	17.5	20.0	8.5
	Crop rotation	Unfertilized	7.5	11.9	9.4	8.4	10.0	4.3	12.8	14.2	4.6
	Crop rotation	Fertilized	18.5	23.2	13.3	11.0	10.3	7.0	14.2	21.5	4.6
Sugar beet	Monoculture	Unfertilized	30.0	13.8	3.1	7.0	5.5	5.5	10.0	20.0	8.0
	Monoculture	Fertilized	66.0	16.7	21.8	8.8	9.5	7.9	16.8	20.8	5.5
	Crop rotation	Unfertilized	21.5	13.7	21.8	11.5	8.0	6.3	20.0	12.0	4.6
	Crop rotation	Fertilized	47.0	19.8	25.7	10.3	8.7	6.2	24.2	20.0	4.0
Maize	Monoculture	Unfertilized	10.0	14.2	28.9	13.0	14.4	2.7	10.6	10.3	3.0
	Monoculture	Fertilized	12.7	6.2	25.7	16.0	19.2	3.0	10.6	12.0	7.0
	Crop rotation	Unfertilized	11.5	7.8	18.3	8.9	7.5	6.2	7.9	11.0	3.0
	Crop rotation	Fertilized	18.5	8.7	27.3	14.4	12.3	7.0	2.3	12.7	8.0

Dehydrogenases and other enzymes were influenced by the seasonal variations, which affect the complex of soil environmental factors. By the end of vegetative phase the dehydrogenase activity is reduced. An interrelationship was established between the dehydrogenase activity and energy of nitrification process, which depended on the activity of a given enzyme and soil fertility. The increase in dehydrogenase activity was the characteristic of winter wheat and sugar beet monocultures. A lower activity of enzyme was characteristic of maize rhizosphere in both, monoculture and crop rotation, which was evident from the inhibitory effect of root excretions on this enzyme. Only the application of organic fertilizers had a positive effect. Maize monoculture have lower dehydrogenase activity than crop rotation.

The plant excretions have a significant effect on the activity of dehydrogenase enzyme. Depending on the character of the given plant species on dehydrogenase activity level will increase or decrease in monoculture.

1.4.2. *Catalase* : The activity of catalase, catalysing the reaction of hydrogen peroxide decomposition into water and molecular oxygen was determined by titrometrical method (Katznelson and Jershov, 1958). Kouprevitsh and Scherbakova (1966) reported that vegetation had a large influence on the catalase activity of soils. Generally soils occupied by tap root plants with deeper penetrating and profusely branched root system have high catalase activity. A positive influence of plants on soil activity was supported by observations of Verona (1952), who found that catalase activity increased noticeably in the soil near the germinating seeds of wheat, maize and alfalfa. The activity of enzyme increased with the application of organic fertilizers into the soil. Tshunderova and Zoubyetz (1968) found that catalase and peroxidase activity increased in monoculture crops than in crop rotation system. We have succeeded in finding this relation only for maize. In wheat and sugar beet crops, the activity of enzyme decreased in monoculture. The level of catalase activity increases in the summer season.

1.4.3. *Polyphenoloxidase* : The activity of polyphenoloxidase enzyme catalysing the phenol oxidation into quinones in the presence of atmospheric oxygen also increased on the fertilized land. However, the lowest level of its activity was noted in the monocultures of winter wheat and sugar beet. It pointed to the fact that decomposition of phenolic substances was significantly more intensive in crop rotation. Because of this, there was a possibility of phenol accumulation in a monoculture.

1.4.4. *Peroxidase* : The activity of peroxidase leads to oxidation of organic soil substances viz., phenols, amines and other compounds in the presence of hydrogen peroxide, also changed depending on the species and conditions of plant growth. The maximum peroxidase activity was found in the rhizosphere of sugar beet and maize in crop rotation and in the control treatments (without fertlizers). The increase in its activity was noticed in monoculture of winter wheat. Results of investigations on the activity of hydrolytic enzymes and oxidoreductases indicated that different plant species influence the synthesis and hydrolysis of organic compounds in soil.

1.4.5. *Proteolytic Enzymes* : Considering the leading role of proteolytic enzymes in the dynamics of assimilated nitrogen forms, we studied the protease activity in 1976-1977 (Table 12). The term 'proteases' denotes a group of enzymes, catalysing the hydrolysis of proteins into amino acids. Nitrogen enters the soil primarily in organic form, mainly as proteins. The proteolytic enzymes break down the proteins in to polypeptides and then to amino acids, which are further decomposed to ammonia, carbonic acid and water. Proteolytic activity of the soils was determined by the intensity of ammonia nitrogen colour.

To determine the protease activity, the method of Galstyan (1965) modified by A.I. Tshunderova was used. The results showed that the highest protease activity was characteristic of plots with organic fertilizers. A close relationship of proteolytic activity with the amount of organic substances in soil was observed by Uchtomskaya (1952). She reported that proteolytic activity increased proportionally to the amount of organic substances applied to the soil.

In our results, proteolytic activity provides did not show that crop rotation had favourable influence on these enzymes, as reported by Tshunderova (1968) and Buragulova *et al.* (1975). Under conditions of our experiment, this phenomenon was observed only in fertilized field, sugar beet and maize rhizosphere. No positive effects were found in winter wheat grown in crop rotation system.

Table 12. Protease activity (mg protein nitrogen g^{-1} soil day^{-1}) in rhizosphere of crop plants, Research Station, Mironovsk

Crop	Cropping system	Fertilization	1976			1977		
			June 17	July 27	Sept.3	June 16	July 27	Sept. 14
Winter wheat	Monoculture	Unfertilized	0.311	0.252	0.308	0.230	0.340	0.410
	Monoculture	Fertilized	0.482	0.336	0.450	0.476	0.370	0.520
	Crop rotation	Unfertilized	0.352	0.224	0.239	0.308	0.360	0.240
	Crop rotation	Fertilized	0.352	0.233	0.357	0.345	0.370	0.520
Sugar beet	Monoculture	Unfertilized	0.331	0.224	0.263	0.214	0.280	0.370
	Monoculture	Fertilized	0.316	0.298	0.357	0.345	0.390	0.390
	Crop rotation	Unfertilized	0.277	0.158	0.293	0.233	0.260	0.390
	Crop rotation	Fertilized	0.432	0.277	0.408	0.308	0.430	0.410
Maize	Monoculture	Unfertilized	0.382	0.168	0.330	0.252	0.300	0.450
	Monoculture	Fertilized	0.412	0.240	0.396	0.326	0.430	0.520
	Crop rotation	Unfertilized	0.321	0.186	0.305	0.233	0.360	0.470
	Crop rotation	Fertilized	0.726	0.261		0214	0.460	0.570

We found that total and maximum protease activity was under maize followed by wheat and minimum in sugar beet. These observations show that the enzymatic activity depended on the influence of higher plants, especially their root systems. Because root hairs excrete active protease into their environment (Kouprevitsh 1954, 1958). It was also found that there was some dependence between the number of microorganisms in soil and protease activity (Balicka and Trzebinski, 1956).

1.4.6. ***Urease*** **:** Results about the dynamics of urease activity (the enzyme catalyzing urea to carbonic acid and ammonia) indicated significant increase in its activity with application of fertilizers (Galstyan, 1961; Gamzikova, 1968). Thus a stable increase in the biological activity of soil including that with urease was observed in the plots located in the manured field. In the control plots of crop rotation, where no fertilizers were applied, the intensity of urea hydrolysis remained identical to monoculture. A relatively high activity of urease in the rhizosphere of monocultures of winter wheat and maize was due to the accumulation of post-harvest residues and roots were more intensive under these conditions.

The dynamics of nutrient substances in soil, specifically ammonical nitrogen (Table 13) and mobile phosphorus and potassium (Table 14) according to the data of Mironovsk Station confirmed our hypothesis. The mineralization of organic substances, accumulation of ammonical and nitrate forms of nitrogen in soil was very high and also present in monoculture crops. In the majority of cases they even exceeded these indices in crops grown in crop rotation. Such phenomenon was also observed in the dynamics of easily available forms of phosphorous and potassium in soil, which excludes the possibility to explain crop yields decrease in monocultures, owing to deficiency of plant nutrients in soil. The investigation of the dynamics of nitrogen, phosphorus and potassium in the above ground parts of winter wheat (Table 15 Mironovsk Station data) shows their higher content in crops grown in crop rotation (except higher P_2O_5 and K_2O content in the grain). It seemed that the presence of some factor reduced the availability/supply of plant nutrients but did not decrease productivity in monoculture due to one-sided soil exhaustion.

Table 13. Dynamics of ammonia and nitrate nitrogen in soil in winter wheat field, Research Station, Mironovsk

Cropping system	Fertilization	Depth (cm)	NH_4^+ mg $100g^{-1}$ soil						NO_3^- mg $100g^{-1}$ soil					
			1975			1977			1975			1977		
			Stem extension	Heading	Ripening	Stem extension	Heading	Ripening	Stem extension	Heading	Ripening	Stem extension	Heading	Ripening
Monoculture	Unfertilized	0–20	1.08	1.51	1.62	1.03	0.68	0.98	1.65	0.33	1.30	0.22	0.39	0.83
	Unfertilized	20–40	0.98	1.30	2.28	0.88	0.60	0.58	0.49	0.40	1.07	0.20	0.28	0.83
	Fertilized	0–20	1.13	1.46	1.76	1.40	0.73	1.48	1.66	0.38	2.38	0.25	0.28	1.10
	Fertilized	20–40	1.06	1.30	1.90	1.15	0.80	0.84	1.05	0.50	1.55	0.32	0.22	0.98
Crop rotation	Unfertilized	0–20	1.54	1.62	1.58	0.48	0.65	0.46	0.80	0.34	0.63	0.38	0.24	0.51
	Unfertilized	20–40	1.18	1.19	1.41	0.48	0.36	0.27	0.64	0.28	0.30	0.49	0.18	0.49
	Fertilized	0–20	1.38	1.50	1.79	0.70	0.38	0.72	0.82	0.58	0.26	0.50	0.39	0.71
	Fertilized	20–40	0.84	1.14	1.84	0.33	0.28	0.65	0.54	0.46	0.23	0.58	0.28	0.50

Table 14. Dynamics of easily-available phosphorus and potassium in soil in winter wheat field, Research Station, Mironovsk

Cropping system	Fertilization	Depth (cm)	K_2O mg $100g^{-1}$ absolutly dry weight of soil						P_2O_5 mg 100 g^{-1} absolutly dry weight of soil					
			1975			1977			1975			1977		
			Stem length	Hea-ding	Ripe-ning	Stem length	Hea-ding	Ripe-ning	Stem length	Hea-ding	Ripe-ning	Stem length	Hea-ding	Ripe-ning
Monoculture	Unfertilized	0–20	13.2	13.0	15.2	11.8	13.9	12.8	15.6	15.1	15.5	12.4	12.7	13.1
	Unfertilized	20–40	10.2	12.2	14.7	9.3	12.3	11.1	13.5	10.7	11.5	12.3	11.2	12.3
	Fertilized	0–20	15.0	13.5	20.7	15.2	15.1	13.8	18.1	14.5	17.7	18.3	19.6	19.4
	Fertilized	20–40	11.5	12.9	15.2	13.0	13.2	11.7	14.6	12.6	11.6	16.7	18.3	18.3
Crop rotation	Unfertilized	0–20	9.0	9.6	10.1	—	9.0	9.4	12.8	12.8	12.0	13.6	13.4	13.3
	Unfertilized	20–40	8.5	10.0	11.5	10.0	9.5	10.1	8.8	11.1	12.8	13.1	13.2	13.0
	Fertilized	0–20	11.7	12.1	12.7	12.2	12.9	10.1	14.8	15.9	16.6	20.7	20.6	20.0
	Fertilized	20–40	10.9	11.2	13.3	10.8	9.6	10.2	12.3	10.6	14.6	18.0	17.3	16.9

Table 15. Dynamics of accumulation of N, P and K in aboveground parts of winter wheat, Research Station, Mironovsk (1997)

Cropping system	Fertilization	Development stage														
		Tillering			Stem extension			Heading			Straw			Grain		
		N	P_2O_5	K_2O	N	P_2O_5	K_2O	N	P_2O_5	K_2O	N	P_2O_5	K_2O	N	P_2O_5	K_2O
Monoculture	Unfertilized	4.51	0.80	1.31	2.48	0.57	0.95	1.46	0.46	0.37	0.45	0.12	0.95	1.99	0.95	0.62
	Fertilized	4.92	1.39	1.54	1.82	0.56	0.93	1.16	0.47	1.15	0.52	0.30	1.25	2.11	0.89	0.65
Crop rotation	Unfertilized	5.34	0.83	1.38	2.87	0.78	1.26	2.21	0.71	1.29	0.49	0.12	0.81	2.26	0.88	0.57
	Fertilized	5.10	1.43	1.40	2.60	0.70	1.04	1.77	0.64	1.67	0.47	0.14	1.27	2.21	1.01	0.57

Our experiments conducted for many years in the monocultures have shown (a) decrease in plant productivity, (b) increase in allelopathic koline activity in the rhizosphere and roots of winter wheat and sugar beet, (c) increase in the number of phytohelminthes, and (d) increase in infestation of winter wheat with stem nematode and that of sugar beet with beet nematodes. In maize crop, there were no differences in koline accumulation in soil between the monocultures and crop rotation, the nematodes population was insignificant and the decrease in production in monoculture was insignificant. When investigating the biological activity and dynamics of nutrient substances, we did not find decreased enzyme activity in monocultural crops.

The nutrients content in the plant shoot did not indicate any deficiency of major plant nutrients in the monoculture. However, the availability and uptake of main nutrients were significantly inhibited. This suggested that supply of plant nutrients to plant was inhibited by increased amount of kolines in monoculture. All these factors which decreases plant productivity could be the reasons of allelopathic soil sickness.

Chapter 3

SOIL SICKNESS : FRUIT TREES

P.A. Morozov

Department of Allelopathy,
Central Botanical Garden, 1, Timirvzskaya Street, Kiev, Ukraine

> *"I perfectly know and many times saw old orchards ... I am deeply convinced that it is useless. Your efforts and costs will not compensate you for the losses involved..."* M. Balabanov, Improvement of old apple orchards (Progressive fruit farming and horticulture, 1908, No. 38).

Study of allelopathic phenomena in fruit orchards is of large theoretical and practical interest, because fruit trees being perennial plants grow in their place for dozens of years. Various excretions of fruit tree come into soil and a significant mass of root residues also decomposes. When apple trees or other crops were grown in the same area for long time, there occured soil sickness. Long ago, gardeners noticed that young trees planted in place of old trees during orchard restoration, do not always grow well and their growth was poor than plantings in non-tree soil.

Soil sickness was one of the oldest and most important problems of horticulture. The problem of repeated use of soil under fruit cultures had been known already in the 17th century. It was described by Pitcher and co-authors (1966) referring to the paper of Worlidge (1968). Blake (1947) informed in the book "Peach culture" (Wildman, 1768), the author observed peach oppression when planting this culture after removal of old peach trees. In the manual of fruit farming (Lucas and Medicus, 1886), the authors paid attention to the fact, that failures were possible, if the soil is not thoroughly prepared to plant young trees instead of rotten trees. Balabanov (1908) did not advise the renovation of old infected orchards, because young trees will be infected with diseases and will dry. Their yield will not recompensate the farmer, for costs and labour involved. Pengerot (1913) underlined that soil under old trees was exhausted, tired, infected with decaying plant residues and that it was difficult to grow healthy plants yielding heavy fruit crops. Janson (1936) reported unsatisfactory growth of fruit plants in nursery, orchards and in plantations along the streets, if the same species was grown in the previous place. As a rule, if a young tree was planted at once in place of an old cut down tree, the young tree do not grow well. He did not support the method of orchard restoration by planting young trees between the old rows, since the interaction between them can lead to detrimental results.

Polishtshuk and Sneshko (1954) ascertained that orchard renovation, i.e. the addition of young trees instead of rotten ones in the old plantation gives bad results. Young trees did not grow, their increase was very small and productivity was low. They describe the results of orchard renovation in Kiev, Ukraine. Old trees were cut down in 1949. In autumn, 150 cm wide and 65 cm deep holes were dug out for 2-year old seedlings (of first grade), which were planted in the spring 1950. During planting, 10-12 kg manure was applied per hole, and then root system was filled up with the upper soil layer. The young trees were watered twice. Trunk circles were scarified and weeds were destroyed. Average increases of the main branches are shown in Table 16 and all seven cultivars appeared to be susceptible to soil sickness. Trees were very weak and looked oppressed. Already in the first year, over 10% of them rot and almost the same losses occurred in the successive years. The root system was very weak. In this case, bad growth and the death of young trees cannot be attributed to (a) low quality of the planting material, (b) planting young trees out of time and (c) insufficient care. The factor causing the failure of orchard repair was in the soil itself.

Table 16. Increase of young trees growing in place after the old apple trees

Cultivar	Year 1950	1951	1952	1953
'Reneta Landsberg'	12.9	19.8	10.8	8.6
'Boiken'	10.6	10.2	8.8	10.0
'Kalvil snežnij'	10.8	16.9	10.0	11.6
'Reneta Simirienko'	14.8	16.6	8.4	10.8
'Titovka letniaja'	9.6	16.4	9.4	11.0
'Pepin shafrannij'	10.6	14.4	10.2	12.4
'Papirovka'	11.0	16.8	10.5	11.9

Due to greater emphasis on horticulture in the last 30-40 years, a lot of information concerning soil sickness in fruit orchards and nurseries had appeared in Germany, England, Belgium, Holland, Italy, France, Bulgaria, USA, Canada, Australia, Japan (Klaus, 1940; Proebsting and Gilmore, 1940; Bronsart, 1949; Koch, 1955; Schander, 1956; Gilmore, 1959; Baxter, 1961. D'Herde and Brande, 1962; Mario, 1962; Marle, 1962; Otto, 1966; Pitcher *et al.*, 1966; Gilles, 1968; Kovatshey, 1970). Soil sickness was also observed in Poland, but it did not present a serious problem, as the orchards were established on new lands (Pieniazek, 1973). A wide geographical spread of soil sickness under fruit cultures showed that its reasons cannot be limited only by nutrient deficiency or by pest infestation and diseases.

In our country also, fruit orchards were established everywhere, on each farm, chiefly on new lands, which were not occupied by fruit cultures. At that time it was recommended to plant trees at wider spacing, for example, 98 plants per hectare in apple orchards. The orchards were kept for 50 years and more. However, now the culture of dwarf and semi-dwarf trees was being done, which offered mechanisation especially at harvest (Budagovski, 1976; Tatarinov and Pavlov, 1976). Besides vigorously growing stocks of trees were planted in rows at high density (Morozov, 1976).

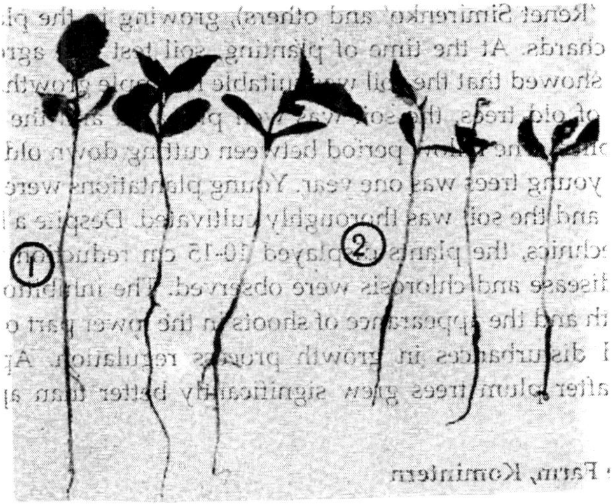

Fig. 5. Effects of extract (1:100) from apple seeds on growth apple seedlings. 1—control, 2—experimental variant.

For intensification of horticulture and location of orchard in the most favourable areas (Popov, 1974), it would be necessary to do fruit plantations in the old orchard lands after cutting down old trees. Despite modern agricultural techniques, the growth inibition and the decline in productivity of young trees grown after cutting down old plantations were often observed. The technology of intensive production of fruit farming envisaged a 5-10-fold increase of plants per unit area (Mouhin, 1976 and Stepankov, 1976). The high density of apple trees will be inductive to soil sickness, the role of allelopathic factor increasing due to the increase in the amount of root residues. Therefore, repeated use of orchard fields under fruit plantations and counteracting the soil sickness will be a serious problem.

1. RESEARCH FARMS

1.1. State Farm, Chkalov

In 1975-1976, young orchards planted after cutting down old plantations were inspected in the southern districts of the Ukraine. At the Crimean experimental station of horticulture and on the Chkalov state farm, we observed a strong suppression of young apple trees (cv. 'Jonatan', 'Jonared', 'Red Delicious', 'Golden Delicious', 'Winter

Banana', 'Renet Simirenko' and others), growing in the place of old apple orchards. At the time of planting, soil test and agrochemical analyses showed that the soil was suitable for apple growth. After the removal of old trees, the soil was well prepared and the fertilizers were applied. The fallow period between cutting down old trees and planting young trees was one year. Young plantations were regularly irrigated and the soil was thoroughly cultivated. Despite a high level of agrotechnics, the plants displayed 10-15 cm reduction in height. Rosette disease and chlorosis were observed. The inhibition in stem tip growth and the appearance of shoots in the lower part of the stem indicated disturbances in growth process regulation. Apple trees planted after plum trees grew significantly better than apple after apple.

1.2. State Farm, Komintern

Apple trees were planted after apple for the third time in the Valley of River Kacha (Apple orchard was first established in 1905). In the summer of 1975, the trees were 2.5-3.0 m tall, with the crown diameter of 2.7-2.9 m. No signs of growth reduction were noted. In this case the success of apple monoculture can be explained by flooding the orchard area during high water in spring. The River Kacha also overflows in summer time after heavy rains. It is supposed that the soil was washed out and plant excretions are removed from the root zone. In our opinion, a more essential role can be played by a temporary anaerobiosis, replaced by normal aeration, which sharply reduce the amount of kolines in soil and changed their chemical activity.

1.3. State Farm, Gardener

A harmful effect of sweet cherries on peach was observed on the State farm "Gardener". It manifested itself in poor growth, drying shoots, poor productivity and failure of trees. Neither a repeated renovation of the young peach orchard, nor fertilizers could overcome this problem. Similar symptoms were observed on the State Farm Chkalov, where the peach was replanted in the place of old peach orchard. When seed- fruit trees were planted after stone-fruit trees, the magnitude of soil sickness became less. Thus, apple after apricot and pear after cherries grew better, than apple after apple or pear after pear.

1.4. Horticulture Research Farm, Melitopol

A strong growth inhibition of cherry planted after cutting down old cherry orchard and when replanting individual fruit trees, the influence of soil sickness was observed at this farm. The influence of soil sickness was appreciable in the year of planting and was displayed upto 2-4 years old and older trees. Soil sickness results in to small pale green leaves on the whole tree or on separate parts of its crown, senescence and drying of shoots and entire branches, a decrease of fruit yield and their quality. Unfortunately, there was no possibility on any farm to compare the repeated fruit plantations with fields, where orchards were established for the first time. In this situation, we had to confine ourselves to visual estimation of tree condition in replanted orchards. But if we compare plantations of different farms, the influence of soil sickness was noticeable.

1.5. Other Farms

For example on State Farm, "Pobeda", we saw a young unirrigated orchard, planted on the field used previously in field crop rotation (southern carbonated chernozem soil). In the second year after planting, the trees showed a 50-60 cm increase and there were no tree failures. In repeated apple plantations, even with irrigation their increase in growth was several-fold smaller.

Long and continuous cultivation of fruit cultures may lead to a situation, where soil will be of no use for orchards. Therefore, repeated attempts to replant apple orchard in the Dendropark Belaja Cerkov, on the land where fruit plants (apples, pears, cherries) grew for over 150 years, have been a complete failure. The trees remained alive, but did not grow and dried after 2-3 years. When planting strawberries and raspberries, the places under the old trees were clearly displayed – the plants there lagged behind in their growth in comparison to others and partially fail.

Among seed-fruit plants, autointolerance was expressed maximum in apple and very low in quince and pear. Among stone-fruit trees, peach makes soil very sick and was very sensitive to replantation. Peach was followed (in the order of decreasing autointolerance) by sweet cherries, cherries, plum, apricot (Bronsart,

1949; Schander, 1956). Difference in autointolerance between the cultivars had not been found.

From berry plants, strawberry (*Fragaria grandiflora* Ehrh.) was very susceptible to a detrimental effect of monoculture. Gooseberry (*Grossularia* Mill.) and currant (*Ribes nigrum* L.) have low sensitivity to repeated cultivation, hence, plantations of these crops can be done many times in one and the same place (Friedrich, 1970).

I. Soil Texture

A display of soil sickness depended on the soil texture. On heavy, less and loamy soils (having a high absorbing ability) and in the regions with insufficient amount of precipitation it may have an influence on tree growth already in the second planting, whereas on light, even poor soils (with a lower absorbing ability), the soil sickness occured significantly later. There were old areas of horticulture, where land had been used under the same fruit cultures for hundreds of years. Thus, on rich loamy soils of the old land region (lower Elba) in Germany, apple and cherries had been grown for already 500 years and only in the recent years signs of soil sickness have appeared there. There were such plots near the Bodensee, which was located on the border between Germany, Switzerland and Austria. Soils resistant to sickness were characterized by a constant and abundant water supply and submergence during floods in the lower reaches of the rivers (Schander, 1956; Friedrich, 1970; Kramer *et al.*, 1975).

2. SOIL SICKNESS

The problem of soil sickness had been discussed in the literature since the last century. There were different explanations of this phenomenon – deficiency of nutrients in soil, especially microelements as a result of their one-sided taking up, soil infestation with nematodes, accumulation in soil of toxic products of plant and microorganism metabolism (Schander, 1956; Börner, 1961). The theory of toxins by A. de Candolle in 1813 (Grümmer, 1957) was the first to explain of the soil sickness. An important contribution to the development of this theory were investigations of Pickering (1903, 1917, 1919), who showed that root excretion of apple inhibits growth of different plants including apple, pear and cherry seedlings. He obtained root excretions by washing out soil (in pots), in which apple seedlings were grown. Water leachates was used for watering tested plants. Pickering came to a conclusion that inhibitory influence of one plant upon another was a usual thing, wide spread in plant world.

Later Janson (1936) suggested that the main cause of soil sickness were root residues, during decomposition of which toxic substances are released. The special important role in autointolerance of fruit plants was ascribed to root residues in nurseries, where plants were very dense and their rooting was not deep, which resulted in accumulation of a large amount of root residues in the upper layer of soil. Proebsting and Gilmore (1940) studied the reasons of failures when renewing peach plantations in California. These authors reported that young trees planted in the place of cutted down peach orchards, were very oppressed and part of them died. The application of agrotechnical methods, fertilizers including microelements and some growth substances gave no positive results. In vegetation experiments, they excluded deficiency of nutrients and pathogenic organisms as a reason of poor growth of trees and showed that in sand culture the peach roots excrete toxic substances. The application of very small root fragments to fresh, not exhausted soil led to the oppression of peach seedlings. Toxic factor, the chemical nature of which was not ascertained, was found in the bark, but not in the wood of roots and was easily extracted by ethanol.

Results of other researchers (Hildebrand, 1945; Patrick, 1955; Collina and Zucconi, 1967) also indicated the presence of growth inhibitors in peach roots. Gilmore (1959) grew annual peach seedlings in pots. Then he cut off their above ground part and after dying and decay of root system he planted peach seedlings into the same soil. Products of root decomposition induced the inhibition of seedling growth. He mentioned that the problem of peach transferring to fertile soils of California was not connected with deficiency of nutrients and infestation of diseases and nematodes.

The presence of toxins in the exhausted soil was confirmed by Fastabend (1955), who found that watering apple seedlings growing in healthy soil with an aqueous extract from exhausted soil lead to the occurrence of symptoms characteristic of soil sickness. The application of very small root fragments of apple to a healthy soil also caused growth inhibition. Schander (1956), mentioned that the theory of toxins was the only acceptable theory, which explained the reasons of soil sickness under fruit cultures. He maintained that the source of toxic substances were root residues and not excretions of living roots and that soil sickness took place only after cutting down old trees (due to release of harmful substances from decaying old root decay). According to his data, soil toxins can be inactivated by soil steaming and by treatment with chemical preparations (carbon bisulphide, chloropicrin and others). He suggested that one of inhibitors coming into the soil during decomposition of apple roots could be phloridzin.

Börner (1959, 1960) had identified many compounds including phloridzin in fallen leaves and in root residues of apple. When roots were added to different soils, phloridzin was detected already after some hours (i.e. it easily diffuses soil). Trials in aqueous culture showed inhibitory effects of phloridzin and its aglicone – phloretin (in concentration of 10^{-3}–10^{-4} M) on the growth of apple seedlings. However, the author considered that the amount of phloridzin coming into the soil is not enough for a direct action on subsequent plants (Börner, 1961).

Berestecki (1970a, 1971) reported that one of the main reasons of unsatisfactory growth of apple plantations in the place of cutted down old apple orchards was the accumulation of toxic substances in soil. Field observations conducted by him showed that shoot extension in young trees of the cvs. 'Kalvil Snezny' and 'Renet Simirenko' planted instead of cut down apple trees on a grey forest podsol and slightly loamy soil was 12-18 cm, whereas shoot elongation on the land, where orchard did not grow before, was 30-51 cm. When the cut down orchard was replanted between the rows the shoot extension reached 26-34 cm. Results of studies on the influence of water extracts from soil upon the growth of apple cuttings indicate that with the age of trees the soil toxicity in the rhizosphere and in the amount of the total mass of roots also increased.

The level of soil toxicity depended also on soil differences and ecological conditions. Thus on grey forest soil, inhibitors were found already under 29-year old trees, whereas not on meadow-chernozem carbonated, slightly loamy soil, only under 70-year old trees. Soil intoxication clearly manifests itself in repeated cultivation of apple in pots. Seedlings planted in soil, in which apple trees were grown during 3 years, died after two weeks. Soil sieving with the aim to separate roots and to dry them in the sun during two days contributed to a decrease in the content of growth inhibitors. After this treatment of soil, the seedlings did not die, but lagged behind in growth than control. Soil toxity also decreased after the sterilization.

The weakening of young tree growth in old orchard plantations was not connected with deterioration of physico-chemical properties of soil and with deficiency of nutrients (Berestecki, 1971). Under conditions of a vegetation bioassay which lasted for 36 days, Berestecki (1970a) studied the influence of decomposition products of root residues of apple and pear on the seedling growth of these tree seedlings. The seedlings grew better in pots with grey forest soil taken beyond the orchard, where fruit plants were not grown before, than on the same soil from under 60-year-old trees. The application of roots to control soil had an insignificant effect on the plant growth, while, strong seedling oppression was observed in the orchard soil. The fact, that a toxic action of root residues manifested itself so strongly in the orchard soil can be explained by a high inactivating ability of the control soil. Besides, decomposition of roots in the orchard soil could lead to formation of intermediate substances with a higher physiological activity, which depends on microflora composition. Soil toxicity under apple was displayed maximum in the root zone and in the rhizosphere (Berestecki, 1974). This indicated that toxicity of orchard soils was related to the metabolism of root system and the accumulation of its excretions.

2.1. Soil Microflora

The role of microorganisms, which decompose apple leaves and roots, when toxins appear in soil, was experimentally shown by Börner (1963). He separated the fungus *Penicilium expansum* from the apple sick soil to infect sterile soil with the addition of bark and autumn apple leaves. In chromatographic analysis, it was found that this fungus produces patulin (a strong phytotoxic substance and an antibiotic) and also phenolic compounds. Many detailed investigations on this aspect were conducted by Berestecki (1972, 1974, 1975). From the soil taken from the rhizosphere under old apple trees, he separated numerous microscopic fungal species from the genera *Penicillium, Aspergillus, Fusarium,* these produced toxins. Metabolites of eight studied spp. completely inhibited the growth of apple, pear, peach sprouts and germination of wheat, radish and alfalfa. Sprouts of fruit trees were found more sensitive to fungal toxins than seeds of annual plants. Inhibition of seed germination stopped at concentrations of 1 : 60 to 1 : 64, while, the concentration of 1 : 512 inhibited the growth of apple and pear seedlings.

Under the influence of root excretions, a specific composition of microflora was formed in the rhizosphere of higher plants (Krasilnikov, 1958). Root and other plant residues also induce the development of certain species of microorganisms, i.e. they have a selective effect on the microflora spp. during the decomposition process. Plants contain antimicrobial substances, which come into the soil as excretions of living or dead plants.

The influence of long cultivation of fruit trees on soil microflora had been shown by several workers. Martin and his co-workers (Martin, 1948, 1948a, Martin *et al.*, 1957; Martin and Pratt, 1958) have found that soil taken from the orchard and beyond its area has different composition of microscopic fungi. Treatment of soil from under citrus plants by steam and chemicals (carbon disulphide, chloropicrin and others) reduced the negative effect.

Wensley (1956) found relationship between the population of some fungal species in soil and reduction of tree growth after replantation of peach orchards. The partial sterilization of exhausted soil with the chemicals changed the fungal microflora composition and improved the peach growth.

Berestecki (1969) ascertained that with senescence of apple tree, the total quantity of microorganisms in the rhizosphere and in the root zone decreased, the correlation between separate microscopic genera changes and the population of pathogenic forms increased. Displacement of microbial equilibrium under apple trees depended on the soil type, it is more pronounced in grey forest podsol soil than on carbonated flood lands.

Hein (1972) confirmed the data of Schander (1956), which indicated that steaming of sick soil weakens its specific negative influence on the growth of apple seedlings. He suggested a direct or indirect participation of bacteria in the occurrence of soil sickness.

The accumulation of phytotoxins, despite the change in qualitative and quantitative, composition of soil microflora disturbed the plant nutrition processes and create conditions suitable for the development of disease agents.

2.2. Nematodes

The growth inhibition and a decline in plant productivity in monoculture also occured due to the build up parasitic nematodes. In the opinion of many authors, nematodes can be one of the reasons of soil sickness under apple, peach and citrus plants (Baines and Clarke, 1952; Swart-Füchtbauer, 1954; Meijneke and Oostenbrink, 1955; Mountain and Boyce, 1958, 1958a). Damage of peach roots by the nematode *Pratylenchus penetrans* lead to formation of necrotic spots arising due to enzymatic hydrolysis of cyanogenic glucoside - amygdalin and release of toxic compounds - benzaldehyde and hydrocyanic acid (Mountain and Patrick, 1959). Not only cellular enzymes of peach can destroy amygdalin, but also nematodes. Börner (1960) indicated, that about 25% of soil sickness cases found in Holland are due to infestation of fruit plant roots with nematodes.

Hein (1972) reported that the application of nematicide (Terracur P) did not improve apple growth, though the parasitic nematodes were found in the sick soil. Picher *et al.* (1966) also showed that in England, the nematodes could not be considered the cause of depression of young apple and cherry trees after old orchard replantations. It seemed that the infestation of roots with nematodes was not a reason, but a result of soil sickness, since weakened plants were unable to resist their invasion.

2.3. Plant nutrients

Supporters of the theory of soil impoverishment (Bronsart, 1949; Kobernuss, 1951) considered that the major reason of soil sickness was the deficiency of microelements (boron, manganese etc.) in soil due to their removal by fruit trees. However many investigators (Fastabend, 1955; Schander, 1956) did not succeed in improving the growth of young plants on the exhausted fields by applying the major and micronutrients. Börner (1960) said that soil always contained indispensable amount of micronutrients and their shortage was displayed very rarely.

It is known in horticultural practice that zinc deficiency in young trees causes rosette disease. This disease occurs most frequently and heavily in apple trees. This disease occurs not only in orchards, but also in nurseries (seedlings) and grafted plants (Tarasov, 1966; 1972, Kanivyetz, 1970). It presented a functional disorder caused by the unfavourable conditions (Ivanov, 1961) and was one of the symptoms of apple soil sickness. It was spread in the old regions of horticulture and may be undoubtedly a consequence of repeated cultivation of apples on the same area.

The question concerning the reasons of soil sickness in fruit orchards had been investigated very little. The fact, that not always soil sickness can be completely removed by applying fertlizers and nematodes indicates the necessity to examine this phenomenon from allelopathic point of view. According to the data of foreign investigators, soil sickness occurs after rooting out of old trees because of their harmful excretions during decomposition of root residues. Only Börner (1961) points to the fact that phloridzin from the autumn apple leaves comes into the soil.

3. OUR STUDIES

It is very important to elucidate the role of soil in allelopathic phenomena and the possibility of koline accumulation under fruit plants. Our aims were (a) to study biological activity and chemical nature of kolines coming into the environment in the metabolism of seed and stone type of fruit plants, (b) to determine the influence of soil properties on the activity of kolines, (c) to develop methods of koline extraction from the soil and (d) their identification. The test fruit trees were different cultivars (mainly in regions) and selective forms of apricot (*Armeniaca vulgaris* Mill.), cherry plum (*Prunus cerasifera* Ehrh.), quince (*Cydonia oblonga* Mill.), cherries (*Cerasus vulgaris* Mill.), pear (*Pyrus domestica* Medik., peach (*Persica vulgaris* Mill.), plum (*Prunus domestica* L.), sweet cherries (*Cerasus avium* (L.) Moench), and apple (*Malus domestica* Borkh.). The species under study belong to the family *Rosaceae*.

Fallen flower petals, ovaries and unripe fruits, leaves as well as cut off shoots and soil samples were 2-3-folds taken from the plot of fruit cultures in (i) Botanical Garden, Kiev (ii) plantations of the Ukrainian Institute of Horticulture, Melitopol, Lvov, Donesk, (iii) Crimea Experimental Stations of Horticulture, State Farms "Komintern", Chkalov etc. in the Crimea district.

3.1. Bioassays

Plant material (except petals and seeds) was chopped into very small pieces and extracted with distilled water for 24 hours at room temperature. The ratio between the weight of fallen leaves or roots and water was 1:20, 1:100, 1:200. To obtain water soluble substances, which are washed out from the fallen leaves, enamel pots were used. Leaves were placed into the pots, kept apart in the Green House in open air. Leakage collected in a saucer under the plant pot was taken after the rain. Aqueous extracts and leakages were used for evaluation of biological and chemical activity. Bioassays proposed by Grodzinsky (1965) – bioassay for seed germinating of radish and bioassay for root growth of cress–were used in our allelopathic investigations. However, we should not confine ourselves only to these two bioassays. When studing specific activity of kolines and the sensitivity of individual species to them (for instance, under conditions of mixed forest cultures, crop rotation and monocultures), we should also use these plants as tests, which are donors and acceptors of kolines. In studies on the elucidation of chemical interaction in plantations of woody plants, the use of bioassay for root growth is methodically suitable. In the tests concerning the studies of allelopathic properties of fruit trees in connection with soil sickness, we used apple, pear, apricot, cherry and other trees simultaneously with radish and cress. Germinated seeds with the rootlet length of 1.0-1.5 cm were selected for the bioassays. Extracts from fallen leaves, roots (1:100) or soil were poured into the test tubes, where germinated seeds were fastened by aluminium foil. Aqueous extracts were added with a nutrient mixture of Helrich (a solution of micronutrients according to Hoagland). To avoid seed drying, a cuvette, containing water was placed for 2-3 days into thermostat, simultaneously with test tubes. After 7-10 days, the inhibition of root growth in the extracts in comparison to the control (nutrient solution) was recorded. The bioassay was conducted with 3 replications (with 15-20 plants per replication).

Seeds of fruit crops must be scarified to hasten their germination. In the summer period it is better to use shoots of fruit trees. Allelopathic activity was indicated by the rate of leaf withering and by the amount of water absorbed. Bioassays with shoots are more comfortable in measured test tubes, which make easier calculation of water consumed by the shoots. The response of shoots to physiologically active substances present in the extracts can be determined by cutting the shoots and observing under the microscope – it reveals the biochemical and structural changes. The bioassay was conducted with 3 replications, each replication containing 10 shoots.

3.2. Pot culture

The soil (grey forest meadow alluvial carbonated humus sand) used for vegetation bioassays was taken from under the trees of apple, pear, cherry, apricot trees etc. Small prices of roots and leaves were applied to soil to soil at 2% (w/w). Then, the germinated seeds of apricot quince, pear and apple were sown into the pots. The control soil was taken beyond the orchard. The bioassay lasted for 60 days. Each replication contained 15-20 plants.

3.3. Identification of Kolines

Physiologically active substances in the fallen leaves and roots were identified using paper and thin-layer chromatography. The material under study was extracted by water and organic solutions. Chlorophyll and other pigments were removed from chromatograms by toluol (Kefeli and Tureckaya, 1966). Several systems of solvents were used to separate the substances: butanol–acetic acid–water (4:1:2, 2; 3:2:95), acetic acid–water (15:85), isopropanol–ammonium–water (80:5:15) and others. Chromatograms were examined under ultrachemiscope to note the spot fluorescence in ultraviolet light (before and after the treatment with ammonium vapours). To display phenol substances, we used 1% ethanol solution of iron chloride, sulphonilic vanilline reagent, Synod sample, Hepfner's reagent. Phloridzin content was estimated by the method of Sarapuu (1965), phenol total – according to Swain and Hillis (Yermakov *et al.*, 1972). The content of organic substances was determined by the method of Simakov and Tsyplenkov (1960).

3.4. Excretion of Kolines

Our studies have shown that excretion of kolines into environment begins from the moment when the seeds are sown into the soil. This was indicated by the following bioassay. Each Petri dish was filled with 5 ml of water and 20 scarified seeds of each fruit tree (apricot, cherry, peach, sweet cherry) were laid out on filter paper. Three days after, these seeds were removed and radish was sown. The excretions of cherries inhibited the germination of radish, sweet

Table 17. Allelopathic activity of water extracts (1:20) from litter and roots of fruit plants

	Koline contents (Relative Coumarin Units)						Root growth (percentage of control)					
Fruit tree spp., cultivar	Seeds	Stems	Petals	Fruits	Leaves	Roots	Seeds	Stems	Petals	Fruits	Leaves	Roots
Apricot, 'Litovchenko'	0.0	151.3	182.5	24.9	560.0	375.1	47.3	4.5	15.0	23.0	3.7	14.5
Quince, 'No 4-8-7'	36.7	147.0	56.0	17.8	224.9	134.4	15.6	18.0	41.5	49.3	9.9	37.6
Cherry plum, 'Lubovskaya'	15.8	132.8	79.3	39.6	176.6	113.7	79.4	19.2	39.3	26.1	17.1	87.4
Cherry, 'Lubskaya'	20.4	479.6	168.7	43.2	545.0	498.5	65.5	0.0	37.2	38.0	9.1	10.3
Pear, 'Lemonka'	45.0	163.2	74.6	101.0	240.3	152.2	49.6	10.9	20.4	42.5	13.0	19.8
Peach, 'Druzba'	3.6	418.5	190.1	46.4	615.0	419.3	29.1	8.5	10.8	11.0	3.6	5.7
Plum, 'Vengerka obyknavennaya'	18.0	122.8	113.0	42.8	296.5	188.6	72.8	2.4	33.9	17.4	6.9	15.2
Sweet cherry, 'Dragana zeltaya'	24.8	500.6	175.9	305.0	416.1	377.0	71.0	3.7	12.7	0.4	12.2	8.6
Apple, 'Jonathan'	31.2	520.0	40.5	28.0	235.2	172.5	80.9	29.2	43.2	21.0	3.7	6.9

cherry, apricot and peach by 35, 24, 25 and 22%, respectively. The inhibitory effect in the bioassay was shown by the extracts from unscarified seeds (Table 17). Radish did not germinate at all in the extract from apricot seeds, whereas, germination in the extract of peach seeds was inhibited by 96.4% in comparison to control. This was explained by the excretion of hydrocyanic acid from the seeds of apricot and peach. In spring, after tree trimming there are shoots and small branches falling under the trees, their water extract has also shown allelopathic activity. The fallen flower petals and fruits were also found to contain kolines. The main sources of kolines in fruit orchard are fallen leaves and roots. In bioassay, the effect of leaf extracts was stronger than root extracts. All the trees have quality different contents of inhibitors (Table 18).

Table 18. Allelopathic activity of water extracts (1:20) from fallen fruits and leaves of apple and cherry

Fruit cultivar	Kolines content, (Relative Coumarin Units)		Growth of cress roots, (% of control)	
	Fruits	Leaves	Fruits	Leaves
Apple				
'Melba'	12.8	320.0	39.8	17.7
'Borovinka obyknawiennaja'	19.3	405.0	58.4	12.9
'Avgustovskoje'	10.7	264.3	13.4	22.4
'Slawa pobieditielam'	16.4	287.4	26.1	25.0
'Donieszta'	10.5	252.5	38.3	23.0
'Pepin litowskij'	17.0	275.6	43.3	20.9
'Kalwil snieznyj'	16.7	300.5	16.2	19.3
'Renet Simirienko'	14.0	288.0	36.7	20.8
Sweet cherry				
'Mliejewskaja czernaja'	234.6	501.5	0.9	17.8
'Jabule'	182.7	400.0	1.3	11.0
'Franz Josef'	359.0	447.3	1.6	10.1
'Roman Oliva'	215.2	380.0	0.0	12.2
'Giedielfinger'	262.8	400.0	0.8	11.8
'Denisiena Żieltaja'	303.0	401.0	0.4	13.9
'Badaksoni'	373.5	501.0	0.0	10.6

Kolines from the fallen leaves are leached by the rain and comes into the soil. This is confirmed by the results of following bioassay. Fallen leaves, (not chopped to small pieces), were placed into the pots (with saucers), at 80 g each. After the rain, the flowing down liquid was collected in the saucer and tested. Rain "extracts" inhibited the root growth of cress; kolines mostly leached from the leaves of cherry and sweet cherry (Table 19).

Table 19. Effect of kolines leached by rain from leaves on root growth of cress (% of control) (1971)

Fruit tree spp., cultivar	Dates of Rain leachate collection		
	October 4	October 5	October 18
Apricot 'Litovchenko'	41.8	34.5	64.4
Cherry 'Lubovskaja'	17.8	17.9	16.9
Pear 'Panna'	47.4	57.8	46.0
Peach 'Poleskij'	57.6	67.2	71.5
Plum 'Wengierka italianskaja'	46.4	44.3	56.1
Sweet cherry 'Jabule'	17.8	15.2	28.6
Apple 'Mc Intosh'	37.2	58.2	44.3

3.4.1. Aqueous extracts

The inhibitory activity of extracts from seeds, petals, leaves and roots was also shown when in bioassay, early seedling growth of fruit trees was studied. We tested the influence of extracts (1:100) from the seeds of apricot, cherry, plum and apple on the growth of apple and apricot seedlings. Apple germinating seeds of the cv. 'Pepin Litovski' with the 1 cm long rootlet were placed into a 20 ml test tube. The rootlet extention was measured after 10 days. In 30-days bioassay with apricot, we used 60 ml test tubes and germinated seeds of the cv. 'Kashchenko No. 30' with 3 cm-long rootlets. Growth of apple seedlings in the extracts was less than control; the root growth was drastically inhibited than the stem growth. In the control treatment, more lateral roots were produced and leaves were larger than apple seedlings grown in the extracts. In apricot seedlings, seed kolines delayed the stem growth by 14-21% and the total root length was 2-3-fold shorter than the control.

Extracts from fallen leaves (1:200) inhibited the root growth of cherry *Cerasus tomentosa* (Thunb.) Wall. The length of roots in the control seedlings was 11.5 cm, whereas that in the extract from apricot, cherry, pears, plums peach and sweet cherry was 3.4 cm (29.5% of control), 3.8 cm (33.0%), 3.6 cm (31.3%), 2.9 cm (25.2%), 3.5 cm (30.4%) and 3.5 cm (30.4%), respectively. Only the extract from plum leaves also decreased the stem growth (88.6%) over the control.

Peach germinated seeds died in the extracts (1:20) from the leaves of apple, peach and apricot. A decrease of koline concentration (1:100 extract) gave seeding height of 9.3 cm, 3.0 and 6.1 cm respectively, than control (9.5 cm), but there was no increase in the primordial root.

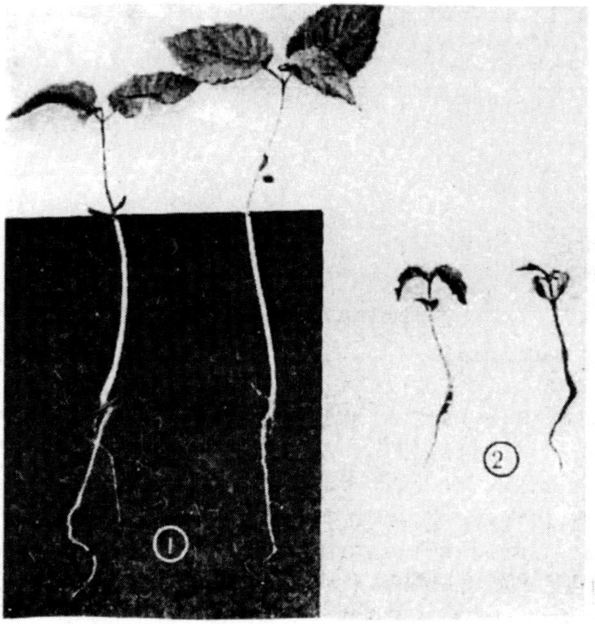

Fig. 6. Growth inhibition of sweet cherry seedlings by extract (1:100) from sweet cherry leaves. 1 – control, 2 – experimental variant.

In 30-days lab bioassay, water soluble substances from the fallen leaves of cherry (Fig. 6) and apple strongly inhibited the growth of cherry seedlings (Table 20). The bioassay showed that substances harmful to fruit plants are contained in the bark and wood of roots (Fig. 7). In the bioassay on the growth of cress roots, bark of apple roots was more toxic than root wood. The root bark extract and root wood extract decreased the stem biomass of cress seedlings by 77.8

and 44% as compared to control. Wood of cherry roots inhibited root growth of cress more strongly than bark. A reverse result was observed, when cherry seedlings were used as test objects.

Fig. 7. Growth inhibition of sweet cherry seedlings by extract (1:100) from sweet cherry roots. 1 – control, 2 – experimental variant.

Table 20. Effect of extracts (1:100) from fallen leaves and roots of sweet cherry and apple on growth of sweet cherry seedlings

Extract	Stem weight (g)	Stem length (cm)	Root length (cm)	Lateral roots (Number)
Control	2.006	12.2	8.5	19
Sweet cherry leaf	0.964	5.8	4.1	5
Apple leaf	0.539	4.6	5.4	0
Sweet cherry Root bark	0.953	8.2	6.1	6
Sweet cherry Root wood	0.570	9.6	5.0	11
Apple root bark	0.447	5.9	7.0	4
Apple root wood	1.20	7.6	5.8	5

Results of the studies on the activity of extracts (1:100) from leaves and roots on the shoot cuttings indicate the presence of harmful substances in the samples. Shoot cuttings of sweet cherry and cherry, apricot and pear withered after 24 hours in these extracts. The maximum whithering was noted in the extract from cherry roots. When cuttings were examined under the microscope, corking of xylem vessels was observed already after 20 hours. Shoots placed in the extracts absorbed significantly less water in comparison to control. For example, the water consumption by sweet cherry cuttings in the first 24 hours was 9.2, 1.3, 2.9, 1.9, 3.4, 2.0 and 3.8 ml in the extracts from cherry roots, sweet cherry roots, apricot roots, cherry leaves, apricot leaves and pear leaves, respectively. This suggests that koline accumulation in soil disturbs the water regime in young plants, when they are planted in old orchard plantations.

A negative influence of soil taken from under old trees on the growth of fruit seedlings was indicated by the following bioassays. The grey forest soil used in the bioassay was taken from under an old apple tree on the "Mour" field in the dendropark "Alexandria". Each pot was filled with 600 g of the soil. Germinating seeds of apple, apricot and quince were sown into this soil. For comparison, these plants were also grown in the soil taken from under an oak tree (from 0-20 cm and 20–40 cm depth). After two months, apple seedlings growing in the soil from under the oak tree (20–40 cm) were 21.6 cm tall; seedlings of apricot and quince were 22.9 and 19.7 cm tall, respecti-vely, while in the soil taken from under the apple tree the seedlings were 14.4, 20.4 and 13.2 cm tall. After the application of apple roots pieces to the soil (2% W/W) taken from under an apple tree (20-40 cm), the inhibition of seedling growth intensified and the seedlings were only 10.8 cm tall. Apple tree grew better in the upper (0-20 cm) than in the lower soil layer taken under the oak tree. The seedlings were 25.0 cm tall, while those growing in the soil under the apple tree were only 16.1 cm tall.

In other bioassay, seedlings of pear were grown in the chernozem carbonated soil taken under pear and apple trees. Seedlings of peach grew in the humic sand soil taken from under the apple and apricot trees. Soil taken beyond the orchard was used as a control. Results of plant height measurements, two months after the beginning of the bioassay showed that pear seedlings grown in the soil taken from under old trees fell behind the control in their growth (Table 21). With the application of apple roots at 2% (w/w) of soil, the magnitude of inhibition of seedling growth increased. Apricot seedlings grew better in the soil under the apple tree than in the soil taken from under the apricot tree. Apple roots reduced the plant growth of pear seedlings to a lesser degree than apricot roots. The application of pear and apple leaves into the soil reduced the growth of pear seedlings and in the treatment with the addition of apricot leaves, the plants died after the appearance of shoots. In this way, phytotoxic substances contained in the roots and in the leaf fall of fruit trees can have allelopathic activity not only in water cultures, but also under soil conditions.

Table 21. Growth of pear and apricot seedlings in soil taken from under old trees

Treatment	Plant height	
	cm	%
Pear		
Control (soil taken outside from beyond the orchard)	7.3	100.0
Soil taken from under pear trees	5.1	69.8
Soil taken from under pear trees + pear roots	3.5	47.9
Soil taken from under apple trees	4.6	63.0
Soil taken from under apple trees + apple roots	3.2	43.8
Apricot		
Control	21.9	100.0
Soil taken from under apple trees	20.2	92.2
Soil taken from under apple trees + apple roots	17.5	79.9
Soil taken from under apricot treess + apricot roots	13.8	63.0

% = per cent of control

A chromatographic analysis of extracts showed that the fallen leaves and roots of fruit crops contain phenolic substances. The chlorogenic acid was found in fallen leaves of all tested species and its quantity in the leaves of quince No. 4-8-7 was 179.0 mg/g of air-dry matter, in the leaves of pear cv. 'Liubimica Klappa' - 75.6, in apricot cv. 'Litovchenko' – 68.4 in peach cv. 'Glory to Kiev' – 42.0, and in plum cv. 'Italyanskaja Vengierka' – 64 mg/g. Leaves of quince were distinguished by a high content of catechines – 62.4-82.3 mg/g of air-dry matter, i.e. 5-10-fold higher than in the leaves of other species. The total content of phenolic compounds in fallen fruits constitute 0.1-0.8 mg/g of fresh weight. Leaves and roots of different pear cultivars were found to have arbutin, whereas, phloridzin was identified in the apple roots and leaves. Thin apple roots (1-3 mm) contained more phlorizin than thick ones (1-2 cm), but the roots contained more phlorizin in the bark than in the wood. For example, in the thin roots of 12-year-old trees of the cv. 'Kalvil sniezny' (rootstock M 2) the amount of phloridzin constituted 15.8% of dry matter, in thick roots – 7.1%, in root bark – 15.4%, in root wood – 1.1%; in the cv. 'Jonatan' it amounted to 12.8%; 6.6%; 18.2%; 2.5%, respectively. Fallen apple leaves contain phloridzin, the cv. 'Doneshta' contained 1.4% phloridzin, 'Kalvil sniezny' – 3.6%, 'Jonatan' – 3.9%, 'Pepin Litovski' – 5.3%, 'Renet Simirenko' – 5.7% of air-dry weight.

To determine the biological activity of phenolic substances, they were extracted from water extracts (1:20) by ethyl acetate. After evaporation of ether, its remains were dissolved in distilled water, the volume of which was equal to that of the initial extract and used bioassay. Ether fractions of extracts from fallen leaves of apricot, quince, cherry plum, cherry, pear, peach, plum, sweet cherry and apple inhibited growth of cress by 56-89% over the control. In bioassay using the chromatograms of ether fractions of leaf and root extracts, we also determined the inhibition zones of cress root growth, corresponding to the location of phenolic substances.

With the help of Institute of Organic Chemistry, Kiev, dealing with natural physiologically active substances in the sick soil, we identified fatty acids and some carbonylic and phenolic acids (Svishtshouk *et.al.*, 1975; Svishtshouk *et al.*, 1977). Grey forest soil (0-40 cm layer) was taken from dendropark "Aleksandria" from under the old apple cv. 'Doneshta' in the field, where fruit apples have been grown since 150 years. Soil samples were placed into glass jars with tight stoppers and mixed with chloroform. Fatty acids of lipid fraction of soil were determined by gas-liquid chromatography on the chromatograph YX-1 (a hard bearer – chromosorb, immobile phase-polydiethyleneglicol adipat; a gas bearer - helium, gas speed - 20 cm^3 for 26 s, temperature of thermal block 233°C), identified them in the form of saturated and unsaturated methyl ethers, after the appearance of methyl fatty acids. Before the analysis, the chloroform was separated by filtration, the soil was air dried, treated with 10% hydrochloric acid (1:5 ratio), washed with distilled water to a neutral reaction, dried again and extracted by a mixture solvents of chloroform - methanol (1:1) to obtain light colour extracts. This extract was fractionated by the method of interphase distribution of solvents. For this purpose, the extract was diuted with distilled water to 5% concentration, the chloroform layer was separately collected, dried by waterless sodium sulphate and concentrated under the water jet vacuum. The excreted brown buttery-looking product – lipids (0.14% of soil taken for analysis) was subjected to saponification by 10-fold amount of aqueous methanol solution of caustic potassium (1:4:1). The saponificated sample was mixed in the reaction flask by a provided mixer and a retroactive refrigerator during 12 hours at 60 °C. The hydrolysis reaction was controlled by adsorbtion chromatography in a thin layer on the plates with diatomite (3x15 cm) in the system of solvents petrolium ether–sulphuric ether–glacial acetic acid (70:3:2). The end of saponification was determined after the disappearance of spots characteristic of initial lipids in ultraviolet light.

Methanol from the reaction mixture was distilled at 50°C; the remains were diluted with double volume of water. Substances, which were not saponified, were thoroughly extracted and separately examined. Water solution of potassium salts of fatty acids was acidified by 0.5 N of sulphuric acid and repeatably extracted by ether. The combined extracts were washed with boiled distilled water until neutral reaction. They were dried by tempered sodium sulphate and the dissolvent was distilled off. As a result we obtained a buttery, hardening product of light-brown colour with the acidic number 119.8. The reaction of etherification of fatty acids contained in the obtained product was performed in a round flask connected to a retroactive refrigerator by treating them with boiling methanol in the presence of sulphuric acid (3% of the mixture volume); absolutely dry methanol was introduced in a 10-fold amount. The end of etherification was determined by a test for free fatty acids (the acidic number not higher than 2). After the end of etherification, the content of the flask was quantitatively transferred into a separating funnel with the use of sulphuric ether and washed with water until a neutral reaction; ether layer was separated by drying the solution by waterless sodium sulphate. Methyl ethers of fatty acids were isolated from a separate organic layer. With the aim to identify undetermined fatty acids, part of the obtained ethers was subjected to hydrogenation by molecular hydrogen on calcium carbonate with a subsequent chromatography on the gas-liquid chromatograph YX-1. As a result of a chromatographic analysis, 33 fatty acids were found in the exhausted soil taken from under the apple tree (Table 22).

Table 22. Volatile fatty acids in sick soil taken from under apple trees

Number	Ratio of carbon atom number : double bonds number	Acid	Content (%)
1	4 : 0	Butyric	Traces
2	6 : 0	Capronic	Traces
3	7 : 0	Enanthoic	Traces
4	8 : 0	Caprylic	0.83
5	9 : 0	Pelargonic	1.29
6	10 : 0	Capric	1.57
7	10 : 1	Caproleic	1.89
8	11 : 1	Unodecenoic	1.39
9	12 : 0	Lauric	2.56
10	12 : 0	2-methyldecanoic	2.03
11	12 : 1	9-dodecenoic	1.69
12	13 : 0	Tridecanoic	2.81
13	14 : 0	Myristic	3.75
14	14 : 0	2-methyltridecanoic	2.32
15	14 : 1	Tetradecenoic	Traces
16	15 : 0	2-methylmyristinoic	2.51
17	15 : 0	1-methylmyristinoic	1,92
18	15 : 0	Pentadecanoic	10.31
19	15 : 1	Pentadecenoic	4.70
20	16 : 0	2-methyldecanoic	2.64
21	16 : 0	Palmitic	4.72
22	17 : 0	2-methylpalmitic	Traces
23	16 : 1	Palmitoleic	5.27
24	17 : 0	Margaric	5.93
25	18 : 0	Stearic	3.81
26	18 : 1	Oleic	6.70
27	18 : 2	Linolic	6.01
28	18 : 3	Linolenoic	8.59
29	19 : 0	Nonadecanoic	2.52
30	20 : 0	Arachidic	2.57
31	20 : 1	Non-identified	2.11
32	21 : 0	Eicosancarbonic	4.87
33	22 : 0	Behenic	2.69

In the soil taken from under an old oak, the content of saponified lipid fraction was 2.6-fold lower than that in the soil from under the apple trees and it was insufficient for identification of fatty acids on chromatograph. Studies on the spread of fatty acids in the soil under the apple tree should be continued to explain the relationships between their quantitative and qualitative content and soil sickness.

Grodzinsky (1965) indicated that lower representatives of aldehydes and ketones have high allelopathic activity. In particular, a phytotoxic substance was benzaldehyde, which was formed due to the break down of amygdalin contained in the root residues of peach. Hence, we analysed the sick soil for the content of carbonylic compounds. Grey forest soil taken from under an old apple and chernozem soil taken from a nursery under 1-year old apple seedlings were used for identification of various phytotoxic substances. The soil was treated with a 10% hydrochloric acid, then consecutively extracted by methanol, a mixture of methanol and chloroform (1:1), and by acetone. The obtained extracts were combined, concentrated and treated with a mixture (1:1) 2 N of caustic soda and acetone. This led to division of liquids into two fractions – acetone (upper layer) and alkaline (lower layer). The acetone fraction was acidified by diluted (1:1) sulphuric acid to pH 3.0 and extracted by chloroform. Chloroform extract was separated in a separating funnel, washed with water until neutral reaction, dried with sodium sulphate and concentrated under the vacuum of a suction pump. As a result we obtained liquid butter-like mass, which presents a mixture of organic substances of soil including carbonylic and phenolic ones. To reveal carbonylic compounds, we used the reaction of their interaction with 2.4 –dinitrophenylhydrazine, as a result of which hydrazones are formed. We also considered the fact, that acetone condensation and formation of ketone and ketoalcohol occur in the alkaline medium. Therefore, the excreted butter-like matter, besides carbonylic compounds, can also contain substances formed in the process of extraction. Under conditions analogical to those of test variant (but without soil), we obtained products of acetone condensation, which were used as standard of soil extract chromatography. Chromatography of the butter-like substance isolated from soil was made in a thin layer of neutral aluminium oxide, with chloroform as solvent. Chromatograms were displayed by 0.4% solution of 2.4-dinitrophenylhydrasine in 2 N of hydrochloric acid.

A carbonyl compound with Rf 0.40 was found in the chernozem soil, whereas that with Rf 0.53 was detected in grey forest soil.

Products of acetone condensation were present in one spot with Rf 0.13. The obtained hydrazons were eluted from the chromatograms and were cleaned in a column with neutral aluminium oxide (15 x 0.5 cm), with benzol and chloroform as solvents and then the temperature of their melting was determined. Hydrazon of carbonyl compound contained in the chernozem soil melted at the temperature of 182–183°C, while that in grey forest soil melted at 145-147°C. Hydrazons of acetone condesation products melted at the temperature of 165-167°C. On the basis of a comparison of the revealed carbonyl compounds with Rfs of standards and data on the melting temperature of their hydrazons. It can be concluded that sick chernozem soil contained isobutyric aldehyde, whereas grey forest soil contains cyclopentanon.

To determine phenolic compounds, liquid buttery substance was applied to the plates with a thin layer of diatomit. Separation was performed in chloroform. Chromatograms were looked through in ultraviolet light and displayed by 1% iron chloride, dinitrogen paranitroaniline, 1% vanilline solution in concentrated hydrochloric acid, dinitrosulpho acid, by 4% solution of 2,4-dinitro-phenyl hydrazine in 2 N hydrochloric acid. On the chromatograms of the extract from the sick soil, one substance was displayed by iron chloride, three substances - by dinitrogen paranitroaniline and three substances by vaniline reagent. The extract from the control soil was found to contain only one phenolic compound, which appeared to be a product of acetone condensation.

Resorcin (Rf 0.14) was identified on the thin-layer chromatograms. Using the method of paper chromatography in the system of n-butanol–acetic acid–water (4:1:5), phloridzin (Rf 0.81), quercitin (Rf 0.71), salicylic and cinnamic acids were identified in the sick soil. Flavin pigment, apparently quercitin was also found as well as a phenolic compound with Rf 0.93, which reacted with di-nitro-para-nitroaniline (scarlet spot) and iron chloride (grey-bluish spot). It is known that iron chloride is a specific reagent on phenols and with trioxy-derivative turns from grey to blue coloration. Localization and spot coloration permit to suggest that this substance is phloretin. According to literature data (Grodzinsky and Grodzinsky, 1973), phloretin Rf in the system of n-butanol–acetic acid–water (4:1:5) is equal to 0.93.

I. Phloridzin

Phloridzin is a growth inhibitor. It can come into the soil from the decay of fallen leaves and root residues. We have found the accumulation of phloridzin in the sick soils of old horticultural regions in south Ukraine (Moroz and Bukolova, 1977). Soil samples, were taken with auger in November 1975, from the orchards of the Crimea Experimental Station of Horticulture, where old trees of the cv. 'Sara Sinap' were cut down the day before. On the State Farm, "Komintern" (Bacchesaray, Crimea district), the soil section was dug out near a 5-years old tree of the cv. 'Renet Simirenko' in the orchard located in the valley of the River Kacha. A young orchard was planted after removal of old plantations. On the Horticulture Eexperimental Farm Melitopol, the soil section over 2 m deep was dug out in May 1976 under the 40-year old apple trees of the cv. 'Rozmarin byely'. Samples were taken from each 10 cm layer upto 180-190 cm. Soil directly from the roots was dug from 30-60 cm depth and analysed separately.

Air-dry soil was sieved through a 1 mm net sieve and small roots were separated. The sieved soil (200 g) was extracted for 24 hours with 80% ethanol (150 ml), then for 1 hour with 96% ethanol (50 ml). The extracts were filtered, combined and evaporated. Dried residue was dissloved in 10 ml of ethanol and chromatographed on M paper. The chromatographs were seen in ultraviolet light before and after ammonium treatment, were examined in bioassays and by treatments of diazo-sulphanilic acid (DSA) and 1% ethanol solution of iron chloride. The chromatograms displayed a substance, which fluorescence in ultraviolet light (scarlet spot) and after spraying with diazo-sulphanilic acid it turns orange, but after spraying with iron chloride, it turns violet. In the system of solvents n-buthanol–acetic acid–water (the component ratio is 4:1:2,2 and 3:2:95) its Rf is equal to 0.85 and 0.30.

To determine the biological activity of the substance, we used the chromatogram obtained when separating the extract from the Crimean Experimental Station soil (40-45 cm depth). We cut chromatogram up zones in 4 x 5 cm size, where this substance was localized, and placed it into a Petri dish, moistening it with distilled water (1 ml) and sowed germinated seeds of cress on it. The control seeds were sown on the stripes of chromatographic paper located below the place of the extract application. After 24 hours on the spot of the studied substance, the root increase of cress was 2-fold lower than that of the control. Based on comparison of the displayed compound with a standard (chemically pure phloridzin) it was identified as phloridzin.

Phloridzin content was 0.9–12.0 μg/g of soil. Under the apple (cv. 'Rozmarin byely') growing on humic soil, the following amount of phloridzin content was found at various depth in soil (Table 22A). The soil (30–60 cm depth) taken directly from the roots of different thickness was found to contain 7.4 μg/g of phloridzin. Under 12-year-old apple tree of the cv. 'Renet Simirenko' (at State Farm, Komintern), the phloridzin content in the same sample was 1.5 μg/g and under 5-year old apple cv. 'Renet Simirenko' at 40-50 cm soil depth was 0.9 μg/g. The meadow alluvial carbonated, slightly loamy soil in the place of the up rooted apple cv. 'Sara Sinap' (Crimean Experimental Station of Horticulture), the phloridzin content at 40–50 cm soil depth was 2.0 μg/g. The phloridzin was not determined at other soil depths.

Table 22A. Phoridizin content at various depth in apple soil

Soil depth (cm)	Phlordizin content (μg/g)
0-10	12.0
20-30	5.0
40-50	2.2
60-70	5.3
80-90	8.8
100-110	2.8

Thus, roots of young trees planted after cutting and removal of old orchards are in the soil environment containing products of metabolism of previous trees, including phytotoxic ones. The presence of phloridzin and other kolines under apple tree is important evidence in favour of the theory of allelopathic soil sickness (Grodzinsky, 1973), which explains the reasons of growth inhibition and reduction of plant productivity in monoculture caused by the accumulation of toxic plant excretions into the soil. To elucidate the influence of soil on biological activity of water-soluble substances contained in the fallen leaves and roots of fruit crops, we conducted the following tests. Soil (humic sand) was poured into filters, at 100 g soil per filter, then 50 ml extract (1:100) was passed through the soil and then distilled water was applied to obtain a filtrate of 50 ml. The extracts and filtrate were examined in bioassay. The results are presented in Table 23.

Table 23. Effect of soil on extract toxicity

Treatment	Growth of cress roots (% of control)	
	In extracts	In soil
Apricot 'Krasnoshchekij'		
Extract from leaves	88.3	-
Extract from root bark	15.4	-
Soil from under apricot + extract from leaves	42.8	60.0
Soil from under apricot + extract from root bark	42.9	72.5
Soil from under apricot + water	79.4	87.2
Control soil + water	101.6	100.0
Pear 'Wiliams letnij'		
Extract from leaves	83.5	-
Extract from root bark	35.2	-
Soil from under pear + extract from leaves	49.9	27.9
Soil from under pear + extract from root bark	90.5	52.3
Soil from under pear + water	67.3	54.6
Control soil + water	76.4	100.0
Sweet cherry 'Franz Josif'		
Extract from leaves	32.2	-
Extract from root bark	67.7	-
Extract from root wood	73.1	-
Soil from under sweet cherry + extract from leaves	32.4	33.8
Soil from under sweet cherry + extract from root bark	49.4	52.4
Soil from under sweet cherry + extract from root wood	58.9	47.6
Soil from under sweet cherry + water	77.9	90.0
Control soil + water	82.5	100.0
Apple 'Rosmarin bielij'		
Extract from leaves	39.5	-
Extract from root bark	16.9	-
Soil from under apple + extract from leaves	30.8	18.9
Soil from under apple + extract from root bark	36.8	38.7
Soil from under apple + water	82.8	81.0
Control soil + water	88.7	100.0

II. Soil Adsorption

Soil adsorbs the organic substances, hence, the extracts toxicity decreased. In particular, cases their toxicity increases – filtrate inhibits the growth of cress more strongly than the initial extract. Because in the process of exchange adsorption, substances with a high inhibitory activity pass from soil to liquid. The total content of organic components in the extracts after passing through soil decreases (Table 24).

Table 24. Carbon content of organic compounds in extracts before and after filtration through soil

Treatment	Carbon content (mg per ml)			
	Apricot	Pear	Sweet cherry	Apple
Extract from leaves	0.75	0.87	0.41	1.39
Filtrate	0.54	0.59	0.27	1.00
Extract from root bark	1.06	0.66	0.31	2.59
Filtrate	0.52	0.30	0.13	2.28

Owing to soil adsorption of kolines, they were accumulated under fruit trees. As seen from Table 23, the increased root growth in the soil where extracts passed through was significantly lower than, when water leached through the soil. Water extracts (1:1) from carbonated soil taken in the rhizosphere also displayed allelopathic activity in the bioassay for root growth of cress (Table 25). Absorption of kolines depends on soil texture and on their chemical nature. Thus, chlorogenic acid was absorbed by weak-humic sand in the smallest amount - from 10 mg of the acid applied to 50 g of soil, 5.6 mg was extracted by water after one hour, while in common chernozem under the same conditions, only 2.2 mg was extracted by water. Carbonated chernozem (from under cherry) was distinguished by the highest absorption ability in relation to chlorogenic acid. A sample of 30 mg was completely bound in soil and at the application of 40 mg, we succeeded in washing out 6.2 mg of the acid.

Table 25. Allelopathic activity of water extracts (1:1) of chernozem carbonate soil from under fruit plants

Fruit tree spp., cultivar	Growth of cress roots, (% of control)
Quince	68.5
Cherry, 'Anadolskaja'	58.9
Pear, 'Bere Bosk'	71.7
Peach, 'Sotchnij'	66.6
Plum, 'Renklod Altana'	70.5
Sweet cherry, 'Dragana Jabule'	57.6
Apple, 'Sari Sinap'	73.0

Phloridzin was adsorbed most strongly by weak-humic sand, though the content of humus in this soil was 2-fold lower than in common and carbonated chernozem. At the application of 10 mg phloridzin to 50 g of soil, 3.3 mg were washed out from weak-humic sand, 7.6 mg – from common chernozem and 8.5 mg of phloridzin – from carbonated chernozem. The soil (weak-humus sand), taken from under apple, absorbed more phloridzin than the control soil taken beyond the orchard. Under aerobic conditions, kolines decompose/ breakdown faster than under anaerobic conditions. Thus, 120 hrs after the application of 10 mg chlorogenic acid to 50 g soil, 1.3 mg were extracted in the treatment under anaerobic conditions and only 0.4 mg of that acid under aerobic conditions. Adsorption ability of soil is very important for chemical interaction of plants, because due to this ability kolines accumulate in root zone.

Agrochemical properties of soil, where fruit cultures are grown for a long time do not change significantly. However, in some cases too much phosphate is applied to the soil. For example, in the Dendropark "Alexandria" Belaja Tserkov, on the fields where orchards were grown for 150 years the content of P_2O_5 in the arable layer of grey forest soil reached 20-39 mg per 100 g of soil, while that on the field with the virgin soil it was 1.8–2.0 mg. With excessive phosphate content in soil, zinc starvation of fruit plants can occur because the zinc change to unavailable form.

To determine of the reasons of soil sickness, the question concerning the taking out of mineral elements with fruit yield is significantly important. In our investigations, we have shown the influence of cultivar, rootstock, growing conditions and agrotechnical measures on carrying nutrition components with the fruits. At the apple yield of 10 t/ha, the following amounts of nutrients are taken up from soil every year: magnesium – 1022 g, calcium – 854 g, iron – 45 g, copper – 3.8 g, manganese – 3.2 g, titanium – 1.8 g, nickel – 0.1 g, molybdenum – 0.05 g. If we consider that a significant amount of mineral nutrients (including micro nutrients) are taken out with the biomass during orchard trimming, we should apply complex fertilizers containing micronutrients to prevent soil exhaustion and provide optimal nutrition of fruit plants.

We have shown that phytotoxic substances accumulate in the soil under fruit trees, the sources of which are fallen leaves, flower petals and dying roots. A large amount of toxins is released after cutting down of trees during the process of root residue decomposition. Koline accumulation in soil is one of the main reasons of unsatisfactory growth of young fruit trees in old orchard plantations. The deficiency of micronutrients because of one-sided removal at tree trimming or cutting down intensifies a negative effect of kolines on plant growth and productivity.

The only method to avoid losses by soil sickness is the planting of young orchards on new areas, not previously used under fruit crops and giving an opportunity to the soil of old orchards to have a rest. However, the application possibilities of this method are very limited. During orchard replantation, it is advisable to fill planting holes with fresh soil carried from the plots under forage crop rotation. A minimum period between old tree cutting down and plantation of new trees should be not less than 3 years. During that period this area should be used for leguminous grasses (clover or alfalfa), because they are more effective in combating soil sickness.

Our investigations have shown that in case of orchard replantation, it is not advised to plant apple after apple, peach after peach, cherry after cherry, sweet cherry after sweet cherry. In the repeated plantations of these cultures, we observed a strong growth inhibition. If the fruit tree cannot be changed, the planting material should be selected on another rootstock. For instance, if the cut down old cherry or sweet cherry trees were grafted on mahaleb cherry, the rootstock of young trees should include wild sweet cherry, in peach – apricot or plum cherry instead of peach, in pear – quince instead of pear, and so on. Simultaneously with searching the ways to eliminate soil sickness, it is recommended to find possibilities to warn about undesirable events. The main approach to solve this problem is to regulate the level and composition of kolines in plant environment.

Chapter 4

SOIL SICKNESS : FLOWER CULTIVATION

N.I. Prutenskaya

Department of Allelopathy,
Central Botanical Garden, 1, Timirvzskaya Street, Kiev, Ukraine

> *"Biological science is close to say its word ..., what plants should be grown together and what plants "hate" each other, owing to excretions of antagonistic phytoncides."* B.P. Tokin, 1974

Soil is the main place where allelopathic interrelations are displayed. Hence, the problem of allelopathic soil sickness due to its significance is becoming main focus of chemical interactions of plants. The allelopathic soil sickness has been established in some crops viz., flax, alfalfa, clover, beet, some fruit crops etc. (Kossovitsh, 1905; Patrick, 1955; Tshunderova and Shulyndin, 1964; Grümmer, 1970). However, flower crops have not been studied in this context. Nevertheless, floricultural practice points to the impossibility to grow most of ornamental plants on the same plot for several years. Therefore, such flower plants as hyacinth, flox, dicentra, gerbera (see appendix for Botanical names) should be replanted after 4-5 years, otherwise their flowers grow smaller, loose their beauty and their yields decreases (Gil *et al.*, 1968). The chrysanthemum developed better and was seldom infected with fungal pathogens on the soils, where it was not grown during the last two years, or was grown on new plots. If there is no possibility to use crop rotation, it is recommended to change the soil in the pots, to frequently lime the soil especially in places where sick plants were found (Shmygin, 1972). After 4-year growth, replantation and rejuvenation of such plants as chrysanthemum, phlox, lupine, delphinium etc. cannot stop the decline in plant productivity. Further use of old field is fruitless (Kiselev, 1952; Kotova, 1974).

The asters cannot be planted in the place of the last year because fusariosis disease will be more intensive, which will kill large number of plants (Petryenko, 1973). To avoid a morbidity infection and decrease in the decorative value of gladiolus, tulip and narcissus, they can be grown in the same place, but not longer than 4-5 years (Gil *et al.*, 1968; Shkolnaya, 1976; Shtshurova, 1976).

Therefore, crop rotation to avoid perennial plant growth in the same place is the indispensable condition for successful growth of flower plants. Scientifically well-founded crop rotation promotes replenishment and better use of plant nutrients, maintains favourable physical properties of soil, prevents spread of weeds, diseases and pests etc. (Syrovatskaya, 1974).

After repeated growth of flower crops, the soil (for these or other reasons) is becoming unattractive for their further successful cultivation. In crop plants one of the reasons for the necessity of crop rotation is the accumulation of toxic products in the soil under them. The available literature directly or indirectly indicates high allelopathic activity of flower plants and a possibility of allelopathic soil sickness under them.

Ballantyne (1962) reported that growth inhibitors accumulated in the soil where chrysanthemum was grown for 5 years. These inhibiting substances were found in the leaves, roots and soil under the growing crops. Growth inhibitors reduced the number of internodes, their length and the number of flower buds. The application of gibberellins (0.1–1000 mg/l) inside the roots slightly decreased, the inhibition of stems, but there was drastic reduction in the number of flowers. Kozel and Tukey (1968) also, found that plants of *Chrysanthemum morifolium* synthesize phytotoxins which was leached released from the leaves and inhibited the growth and development of this specie. Besides, Tukey (1969) confirmed, that chrysanthemum cannot grow on the same place during several years owing to accumulation of inhibitors in soil.

Soil sickness occurs also in begonia monoculture and is displayed in delayed growth, browning of leaves, root dying off and withering of plants (D'Herde *et al.*, 1961). Grodzinsky (1965) indicated that dry fruit-peel of balsam has high content of germination inhibitors. The pericarps of poppy and pink had similar properties.

Some flower crops (purslane, pink, sweet pea and others) contains the endogenous growth inhibitors (Mitsuhashi and Shibaoka, 1965). However, no information is available about their possible participation in chemical plant interactions and their role in soil sickness.

Tukey (1966) reported about the washing out/leaching by rain of various compounds from the above-ground part of plants. The substances leaching from the leaves can accumulate in soil. About 15-19 amino acids, 8–14 organic acids, as well as glucose, fructose, sucrose and other compounds can be washed out from the above-ground parts of chrysanthemum and pink.

Many flower plants are rich in essential oils, which can delay seed germination to a significant degree, inhibit growth of roots and above-ground parts of plants and decrease the intensity of photosynthesis, respiration and other processes. Coupin (1911) reported that volatile excretions from the violet, Alexandria senna/ marjoram, flox, pink and other flower crops delayed the seed germination, seedling growth and injured the leaves of tested objects. Kimmel (1969) informed that plants of the families *Ranunculaceae, Iridaceae, Liliaceae* and *Asteraceae* contained high concentration of volatile inhibitors of germination and growth. The bioassays have shown that volatile excretions of flower crops can penetrate into the soil. Therefore in many cases, allelopathic soil sickness can occur due to accumulation of toxic products under flower plants and have direct inhibitory influence of kolines on the growth and development of these or other species.

It is known that continuous growing for many years of this or other crops results in the impoverishment of microflora population (Grümmer, 1957). Plant excretions and decay products and metabolites of some soil microflora suppresses the development of these or other micro-organism species. A reduction in number of microbe-antagonist leads to favourable conditions for the development of phytopathogenic microflora. On the other hand, allelopathic substances can directly stimulate and depress the development of some pathogenic organisms.

In particular, inconspicuous-looking marigold planted between the rows of gladiolus controlled the to nematodes in tuber bulbs (Yareshko, 1973). Aster planted near the onion feel excellent – grow well, have abundant flowers and are healthy (Popova, 1973). When the of conifer plant needles are incorporated in plot of gladiolus beds, no disease signs were noticed in plants (Andreyev, 1972).

Infusions from a fresh plant matter and dry residues many plant species controls the aphids and ticks on flower crops (Kapa, 1976). Smirnova (1969) found that when the delphinium and gladiolus were grown together, the latter was not infected by *Fusarium oxysporum*. Based on field trials and laboratory investigations, be concluded that defence response of delphinium is carried out by allelopathic mode, i.e. through direct influence of delphinium on the disease agent. Hence, the flower crops represent great scope for investigation of their allelopathic activity in relation to soil sickness under them.

1. FLOWER CULTIVATION

Flower plants have been introduced into cultivation comparatively recently and they have preserved many of their "wild" features including allelopathic activity and tolerance. Flower plants are of different origin, hence, their allelopathic properties depend on it. Selection of flower plants is done to obtain a more decorativeness, resistance to diseases or to unfavourable conditions, however, their ability to allelopathically influence other species and soil sickness have never been considered, hence, not investigated.

The investigation of allelopatic soil sickness under selected flower cultures was started from search for allelopathic compounds in plants themselves and in their residues with simultaneous attempts to find kolines in soil. The bioassays have made it possible to find the presence of toxic products in living plants (roots, stems and leaves) and in their residues (Table 26). The inhibitory effects of identified kolines depends on a number of factors :

(A) **Determining their activity and amount in extract/leachate** (Table 27). The inhibi-tory effect drastically decreased with reduction in the number of physiologically active substances, i.e. with increase of dilution. At 1:10000 dilution, the inhibition was changed to stimulation (radish seed germination data) due to dose dependence effect (when the amount of substance decreases, it changes the inhibition into stimulation).

(B) **Sensitivity of test-objects**. In our experiments, rye roots appeared the best indicators of kolines, whereas, radish seeds were the least sensitive.

(C) **Source of kolines** itself. Our bioassay data clearly showed that Korean chrysanthemum 'Leto' (with red flowers) was more inhibitory than the chrysanthemum cv. 'Tshemshushnaya' (with white flowers).

Table 26. Occurrence of kolines in Korean chrysanthemum and China aster (growth of cress roots, % of control; plant matter to water ratio 1:10 w/v)

Treatments	Live plant		Plant residues	
	Shoot	Roots	Shoot	Roots
Chrysanthemum				
'Tshemtshushnaja' (White flowers)	20.4	66.4	33.0	77.0
'Leto' (Red flowers)	14.6	38.2	24.0	33.0
China aster - dwarf queen group				
White	10.9	18.3	0	]10.1
Pink	8.2	8.9	0	5.6
Red	6.2	5.7	0	4.0
Violet	4.7	5.2	0	7.8

Table 27. Biological activity of water soluble compounds from dry residues of chrysanthemum (% of control), 1974

Ratio of plant resistants water	Radish (Germination)	Root growth				
		Cress	Millet	Wheat	Rye	Barley
Chrysanthmum 'Leto' (Red flowers)						
1:10	18.4	8.4	2.7	7.5	6.2	6.2
1:100	59.0	50.0	46.0	45.3	27.5	62.1
1:1000	100.0	75.0	65.0	75.0	42.5,	72.4
1:10 000	110.0	85.0	84.0	84.5	47.3	78.0
1:100 000	100.8	87.0	89.0	93.0	71.4	88.1
1:1000 000	100.0	95.0	90.0	96.5	84.2	99.1

Chrysanthmum 'Tshemtshushnaja' (White flowers)						
1:10	18.4	9.2	30.0	8.0	9.8	32.1
1:100	82.0	39.0	96.0	52.5	35.0	53.4
1:1000	101.0	110.0	96.0	78.0	50.0	57.1
1:10 000	111.0	107.0	-	80.5	62.0	65.4
1:100 000	120.0	100.0	-	82.5	64.5	93.1
1:1000 000	114.0	100.0	-	88.5	75.3	74.2

Analogical data were obtained with the residues of Chinese aster from the group of dwarf queen (Table 28). In this case, large differences were found in the allelopathic activity between the cultivars.

Table 28. Activity of water soluble compounds from plant residues of China aster from dwarf queen group (percentage of control), 1977

China aster	Solution [ratio of plant residues to water (w/v)]	Bioassay criterion		
		Germination of radish seeds	Growth of cress roots	Growth of wheat roots
Red	1:10	12.0	0.0	0.0
	1:100	88.0	29.3	28.1
	1:1000	108.0	51.0	91.0
	1:10 000	103.0	101.0	82.0
White	1:10	10.0	0.0	0.0
	1:100	94.0	7.7	21.2
	1:1000	96.0	53.0	93.4
	1:10 000	105.0	87.0	90.0
Pink	1:10	10.0	0.0	0.0
	1:100	90.0	13.9	36.3
	1:1000	98.0	87.5	95.0
	1:10 000	154.0	136.0	96.0
Violet	1:10	10.0	0.0	2.0
	1:100	105.0	26.9	35.9
	1:1000	100.0	94.3	91.2
	1:10 000	112.0	135.2	105.7

Our bioassays experiments under laboratory conditions had shown the presence of phytotoxic substances in the flower plant residues. However, toxins found in bioassays under laboratory conditions always did not show allelopathic effects under natural conditions. Because different factors may have neutralizing role viz. : buffer activity of soil, activity of microflora and other biotic and abiotic factors. Besides, a response of the plant itself to its own kolines will be undoubtedly different than the test plant response. Therefore, when estimating the degree of allelopathic effect, it is necessary to make experimental conditions similar to natural ones.

1.1. Chrysanthemum

Vegetation trials done in sand and soil cultures showed that plant residues of Korean chrysanthemum preserve their high allelopathic activity not only in short bioassay, but also in longer trials with the full plants growth cycle (Proutenskaya, 1975). The plant residues of chrysanthemum incorporated into the sand decreased the process of root formation in cuttings of this species. At the early developmental stages, Korean chrysanthemum (both cultivars) displays an active autointolerance to their kolines (Table 29).

Table 29. Effect of chrysanthemum residues on rooting their cuttings in vegetative experiment in sand cultures, 1974

Plant residues applied, (g per kg sand)	Number of roots at 30 days (mean from 50)	Root length (mm)
Cv. 'Leto'		
50.0	6.0	0.9
10.0	9.1	2.6
5.0	13.8	9.6
2.5	17.6	9.7
1.0	34.0	9.8
0.1	45.5	21.4
control	49.0	20.1
Cv. 'Tshemtshushnaja'		
50.0	7.8	1.2
10.0	10.3	1.9
5.0	11.5	2.9
2.5	12.0	5.9
1.0	16.8	5.5
0.1	24.8	17.5
control	23.0	15.9

In vegetation experiment in soil cultures, chrysanthemum residues had significant activity on the growth and development of plants (Table 30). It should be mentioned that during the first month after rooting of cuttings no special differences were found between the separate treatments of the trial. Allelopathic effects become visible during the second month. The degree of inhibition intensified with time and at full bloom, the plant height in the treatments with chrysanthemum cv. 'Leto' with maximum dose of plant residues was 35.7 cm, whereas that of the control was 51.8 cm. In the experimental plants, there was significant reduction in (a) the length of internodes, (b) leaf size, (c) quantity of flower buds and (d) mass of above-ground organs.

Table 30. Effect of chrysanthemum residues on growth and development of this species plants in vegetative experiment in soil culture, 1973

Plant residues applied, (g per kg soil)	Plant (cm)	Inter-node length (cm)	Leaf length (cm)	Leaves width (cm)	Number of buds, pcs	Fresh weight per plant (g)	Dry weight per plant (g)
Cv. 'Leto'							
Control	51.8	2.15	8.5	5.8	9.2	27.5	10.8
10.0	35.7	1.40	5.7	3.7	5.7	11.6	5.5
1.0	49.0	2.01	7.5	5.2	8.5	22.3	9.9
0.1	49.9	2.08	8.0	5.4	10.5	25.8	10.4
Cv. 'Tshemtshushnaja'							
Control	37.7	1.79	8.2	14.0	14.0	14.9	6.4
10.0	36.7	1.62	8.1	13.0	13.0	13.6	6.0
1.0	40.7	1.91	8.7	13.0	13.0	19.2	6.7
0.1	39.7	1.79	8.3	17.0	17.0	15.5	6.5

Chrysanthemum cv. 'Tshemshushnaya' was more tolerant to its kolines. A stimulating effect was observed with minimal amount of its residues to the soil, whereas at maximum dose, slight suppression of the plant growth and development was noted.

In further bioassays chrysanthemum was grown with chickweed plant (a common weed used as an acceptor). The chickweed experienced the largest depression, when it was grown together with the chrysanthemum cv. 'Leto' (Table 31).

Table 31. Effect of two cultivars of chrysanthemum on chickweed, 1973

Plant residues applied (g per kg soil)	Dry mass / plant (g)
Control	5.61
Cv. Tshemtshushnaja	
(g per kg)	
10.0	4.46
1.0	4.99
0.1	6.60
Cv. Leto	
10.0	3.85
1.0	4.05
0.1	6.44
Chickweed sown with chrysanthemum *'Leto'*	3.57
Chickweed sown with chrysanthemum *'Tshemtshushnaja'*	4.91

Therefore, the application of plant residues reduced the growth and development of same or other species, hence, this means that plant residues remaining in soil or on its surface can become a real cause of allelopathic soil sickness. Besides, kolines released during the decay of plant residues and accumulation in soil can control the weed spread. There is a close correlation between allelopathic activity and the degree of soil sickness.

Our investigations of the toxicity of soil samples, taken from under chrysanthemum and aster, were carried out with the use of different modifications of the bioassay method. For this purpose we analysed the aqueous extracts (10 : 1.5), volatile substances (3 g of moistened soil per 50 cm^3 Petri dish) and the soil itself (modified Neubauer method). According to the obtained data (Table 32) the bioassays are quite effective in determining the soil toxicity.

Generalizing the results of bioassays, the largest amount of kolines is contained in the soil from under the chrysanthemum cv. 'Leto' and dwarf queen aster with the flowers coloured pinky and red. As a rule, volatile excretions of sick soil have a stimulatory effect.

Table 32. Estimation of toxicity of sick soil (root length (mm) of test plant)

Test plant/ Treatment	Control	Soil from under chrysanthemum		Soil from under Aster		
		'Leto'	*'Tshemts-hushnaja'*	White	Rose	Red
		Root length (mm)				
Cress						
Soil	23.4	20.7	23.0	22.0	22.5	22.2
Leachates	18.8	17.0	18.0	18.4	15.4	18.2
Volatiles	27.8	26.0	26.4	24.8	24.8	26.9
Wheat						
Soil	42.0	41.5	43.0	44.8	50.7	55.7
Leachates	30.7	30.8	30.9	25.3	29.4	29.7
Volatiles	36.9	36.9	38.4	54.3	36.8	40.0
Rye						
Soil	52.3	30.0	47.8	46.0	40.0	42.0
Leachates	39.8	33.3	32.5	31.7	32.1	29.5
Volatiles	23.5	33.5	36.9	28.5	35.0	27.0
Barley						
Soil	55.3	45.5	57.5	65.7	56.3	50.3
Leachates	18.0	22.8	25.5	22.9	22.6	20.5
Volatiles	18.9	17.5	23.1	13.9	19.7	29.7
Millet						
Soil	22.0	18.0	19.5	15.9	21.0	16.0
Leachates	11.2	13.2	13.3	13.2	12.7	16.5
Volatiles	14.9	23.5	23.5	21.2	17.2	22.5

We have conducted longer lasting bioassays with the sick soil. Air-dry soil (100 g) was placed into porcelain wares and 10 germinated rye seeds were planted per ware. After 10 days, measurements were made. Rye clearly responsed to the presence of kolines in the soil and very significant changes took place in the root (the inhibition was 70%, Table 33). In this manner, testing of the sick soil (for both short or long time) offers the possibility to reveal soil toxicity. Although we have obtained some positive results, but we should continue searching for more effective bioassay to know the content of kolines in soil.

Table 33. Allelopathic activity of soil from under flower plants (% of control)

Treatment (Source of soil)	Wheat			Rye			Barley		
	Plant height	Shoot weight	Root weight	Plant height	Shoot weight	Root weight	Plant height	Shoot weight	Root weight
Chrysanthemum *'Leto'*	115	98	76	82	80	50	73	85	82
Chrysanthemum *'Tshemtshushnaja'*	150	117	78	88	87	55	96	113	122
Aster, white	107	91	70	80	85	30	82	96	36
Aster, pink	93	74	66	53	56	30	99	100	113
Aster, red	96	95	97	82	77	30	91	94	59
Aster, violet	94	99	51	86	95	48	97	126	80

We have also conducted tests under vegetation conditions using the sick soil. The aim of the investigations described below was to model the natural conditions to a maximum. Besides that, we also tried to find how the mineral elements (nitrogen, phosphorus, potassium, iron) influence the change in the sick soil toxicity. In the tests, we used a control and sick soils (from the plot growing flower plants for 2 years). Mineral elements were not applied in the first series of the conducted tests; Helrich nutrient solution (according to methods of tests in soil cultures) was applied in the second series; in the third series, peat and humus were applied as organic fertilizers along with the mineral element. Chrysanthemum plants were planted as rooted cuttings, while asters - as seedlings. The bioassays were conducted in 4 replications. The observation data concerning growth and development of tested plants are presented in Tables 34-35. The data indicate that allelopathic soil sickness under the flower cultures.

In Korean chrysanthemum, the sick soil exerts more allelopathic effect on test rooted cuttings in the initial stage of development, i.e. immediately after their planting. With the application of fertilizers, the allelopathic effects were more evident. The height of chrysanthemum cv. 'Leto' in the control and sick soils without fertilizers was 54.0 and 53.0 cm and the biomass per plant was 5.90 and 4.23 g. At a full rate of mineral and organic fertilizers the height of control plants was 72.0 cm, the biomass – 12.40 g, whereas the height of plants grown in the sick soil was 64.0 cm, and the biomass was 7.08 g (Table 34).

Table 34. Effect of fertilization on changes in toxicity of sick soil from under chrysanthemum 'Leto', 1976

Treatments	Plants height (cm)																Dry wt. of plant (g)
	May 17	May 24	May 31	June 7	June 14	June 21	June 28	July 5	July 12	July 19	July 26	Aug 2	Aug 9	Aug 16	Aug 23	Aug 30	
Control soil																	
Unfertilized	8.3	10.3	13.6	20.3	23.3	26.6	33.3	34.6	37.3	40.0	41.0	42.6	46.6	50.6	53.6	54.0	5.90
NPK	10.0	14.0	17.3	27.6	28.0	32.6	36.4	43.3	44.6	46.0	47.3	49.0	52.6	54.6	55.0	56.0	12.08
NPK+manure	12.0	16.0	20.3	27.6	35.0	42.0	47.0	53.6	57.6	59.0	59.6	61.0	64.0	68.3	70.3	70.3	12.40
Sick soil																	
Unfertilized	7.8	8.3	11.0	14.3	21.3	24.3	28.6	34.0	36.2	42.0	46.3	48.3	50.3	51.6	52.6	53.0	4.23
NPK	8.3	10.0	12.3	18.3	20.6	24.3	27.0	30.6	31.6	37.0	40.3	46.3	48.0	52.3	54.0	55.0	7.20
NPK+manure	8.6	11.6	15.6	24.3	29.3	34.6	41.6	47.0	49.3	53.0	55.0	57.0	58.0	62.3	63.3	64.0	7.08

Table 35. Effect of fertilization on changes in toxicity of sick soil from under chrysanthemum 'Tshemtshushnaja', 1976

Treatments	Plants height (cm)																Dry wt. of plant g
	May 17	May 24	May 31	June 7	June 14	June 21	June 28	July 5	July 12	July 19	July 26	Aug 2	Aug 9	Aug 16	Aug 23	Aug 30	
Control soil																	
Unfertilized	7.6	8.6	11.6	16.6	23.3	23.6	27.0	30.0	33.0	36.0	40.6	46.0	50.0	54.0	58.0	60.6	7.16
NPK	6.3	7.0	10.6	17.0	28.0	28.0	34.6	39.0	43.0	46.0	47.6	50.3	54.0	57.0	60.0	63.0	10.33
NPK+manure	7.0	9.0	12.0	20.0	35.0	35.0	36.0	44.0	48.0	50.3	52.0	56.0	58.0	60.0	64.0	66.0	15.66
Sick soil																	
Unfertilized	6.8	8.3	10.6	18.0	21.3	26.3	30.3	36.0	39.6	41.6	43.0	46.2	50.0	52.4	58.0	60.0	4.90
NPK	5.3	7.6	10.6	18.0	22.3	26.6	33.0	37.0	41.6	43.6	45.3	48.3	50.4	52.1	57.1	59.0	6.16
NPK+manure	6.0	6.6	9.0	13.0	16.6	19.0	23.0	26.0	28.3	32.6	36.6	40.6	44.1	48.3	52.0	55.0	10.80

Similar results were obtained with another chrysanthemum cv. 'Tshemshushnaya'. In contrast to the preceding trials with Korean chrysanthemum (described above), no differences were found between these two cultivars, although they were very distinct in the bioassays and other investigations (Table 35).

These results show that high allelopathic activity was peculiar to Korean chrysanthemum. The toxic effect of plant excretions were displayed not in short time testing, but also in long term trials with entire plants. The inhibitory activity of both plant excretions and soil from under chrysanthemum shows the impossibility to grow this culture on the same area for a long time. It was quite obvious that in a long term growing of this plant, accumulates highly toxic substances in the soil, which leads to allelopathic soil sickness.

1.2. Aster

Still more convincing data were obtained in growing of asters on the sick soil. Difference between the control and tested treatments was large at the end of experiment. Plants on the sick soil looked weak and some of them died. Differences in the dry matter between the control and tested plants, showed the presence of allelopathic effects. The mass of plants growing in sick soil in the presence of fertilizers was 2-fold less than control under the same conditions (fertilizers, moisture, light, etc.) (Tables 36–38). Therefore, we are convinced with the presence of soil sickness, the application of fertilizers (mineral and organic) instead of decreasing allelopathic effect increased. Under natural conditions, we have attempted to increase the occurrence of allelopathic soil sickness by growing aster in one of the plots for 4 years. During the first 2 years, the plants did not lose their decorativeness, but in the third year, their germination was very sparse (less than 10%). As per literature, aster can be grown after different plants, but not on the place where it was grown in the last year due to the infection by pathogenic organisms (fusariosis).

These results clearly indicated that the soil itself from under flower cultures, aster and chrysanthemum (without additional source of kolines) was toxic to them. The vulnerability to diseases occurs very soon as a secondary process resulting from the fact, that plants on sick soil are very weak and therefore are easily invaded by phytopathogenic microflora. We do not deny that plant excretions themselves are source of nutrition to many causative agents of plant diseases. Besides, allelopathic substances can change the micro-biological balance due to suppression of beneficial microflora. It is impossibile to grow flower plants on the same plot, the necessity to change crop is determined by many factors including accumulation of allelopathically active substances in soil.

Table 36. Effect of fertilization on changes in sick soil toxicity from under white China aster from dwarf queen group, 1976

Treatment	Plant height (cm)							Dry weight per plant (g)
	July 21	July 28	Aug. 2	Aug. 9	Aug.16	Aug. 23	Aug. 30	
Control soil								
Unfertilized	4.0	8.0	11.0	12.0	13.0	13.6	14.5	4.17
NPK	6.3	12.6	13.6	15.0	15.3	17.0	17.0	7.03
NPK+manure	6.6	10.0	15.6	16.0	17.3	17.6	18.6	5.00
Sick soil								
Unfertilized	5.0	7.0	12.0	13.3	13.6	13.6	14.0	3.90
NPK	5.3	7.6	12.0	14.0	14.3	14.3	14.3	3.26
NPK+manure	5.5	10.0	13.0	14.0	14.5	14.5	14.5	2.55

Table 37. Effect of fertilization on changes in sick soil toxicity from under pink China aster from dwarf queen group, 1976

Treatments	Plant length (cm)							Dry weight per plant (g)
	July 21	July 28	Aug. 2	Aug. 9	Aug. 16	Aug. 23	Aug. 30	
Control soil								
Unfertilized	7.3	9.3	15.3	17.0	17.6	18.6	18.6	3.56
NPK	6.6	10.0	14.3	16.0	19.0	20.0	23.6	3.58
NPK+manure	8.6	12.6	`8.1	19.6	20.3	20.6	20.6	5.00
Sick soil								
Unfertilized	4.0	8.0	10.0	11.6	12.3	12.6	13.0	1.76
NPK	6.3	8.3	9.3	9.3	9.3	10.0	13.0	0.70
NPK+manure	5.6	6.2	7.0	7.6	9.0	9.0	12.0	2.30

Table 38. Effect of fertilization on changes in sick soil toxicity from under red China aster from dwarf queen group, 1976

Treatments	Plant length (cm)							Dry weight of plant (g)
	July 21	July 28	Aug. 2	Aug. 9	Aug. 16	Aug. 23	Aug. 30	
Control soil								
Unfertilized	7.3	8.3	9.6	10.3	10.3	10.3	12.5	3.33
NPK	6.0	6.6	9.3	9.6	11.0	13.0	13.3	3.43
NPK+manure	6.6	11.3	14.0	16.6	18.0	19.3	19.3	3.25
Sick soil								
Unfertilized	4.3	6.3	8.3	10.0	10.6	12.3	12.5	1.42
NPK	4.3	4.6	5.6	7.3	8.5	9.0	11.5	2.56
NPK+manure	5.6	7.6	9.3	10.0	11.6	13.3	13.3	2.80

These results clearly indicated that the soil itself from under flower cultures, aster and chrysanthemum (without additional source of kolines) was toxic to them. The vulnerability to diseases occurs very soon as a secondary process resulting from the fact, that plants on sick soil are very weak and therefore are easily invaded by phytopathogenic microflora. We do not deny that plant excretions themselves are source of nutrition to many causative agents of plant diseases. Besides, allelopathic substances can change the microbiological balance due to suppression of beneficial microflora. It is impossibile to grow flower plants on the same plot, the necessity to change crop is determined by many factors including accumulation of allelopathically active substances in soil.

The investigation of allelopathic soil sickness under any test plant is impossible without identification of those allelopathic substances, which are contained in the plant itself and which accumulate in soil.

2. KOLINES

In studies to identify flower plant kolines, the main attention was focussed on phenolic compounds because they have major role in allelopathic effects in both, natural and artificial phytocenoses. This group of substances are wide spread in plant world, has a high biological activity and takes an active part in growth processes. Phenolic compounds in comparison to other substances are part of plant excretions (amino acids, organic acids, sucrose), under natural conditions are relatively resistant to the metabolizing through microflora. In studying allelopathic properties of flower plants, we tried to know the correlation between inhibitory activity, on the one hand and phenolic accumulation on the other.

A simultaneous determination of allelopathic activity of plant excretions and a quantitative content of phenolic substances has shown that out of 18 cases of maximum content of growth inhibitors in above-ground parts and roots in soil from under the test plants in 10 cases, it was correlated with high content of growth inhibitors and with flavonol and *orto*-diphenol accumulation.

2.1. Determination of Kolines

The use of complex biochromatographic methods helped in determining the chemical nature of flower plant kolines. The studies were carried out according to the following scheme. A sample of plant material (3 g) was subjected to a fine extraction with 70% ethanol 26°C. The obtained extracts were combined, vapourized and the remains were dissolved in 10 ml of 70% ethanol. The obtained extract was applied at 0.1, 0.250 and 0.500 ml per spot on a sheet of chromatographic paper. The separation of phenol compounds was performed in acid systems of solvents (n-buthanol–acetic acid–water 3:2:95; n-buthanol–acetic acid–water 40:12:28; 15% acetic acid) to avoid their oxidation. Growth inhibition was displayed as usually. At the beginning, the chromatograms were looked through in day light and in ultraviolet light in the atmosphere of ammonia or without it to find phenolic compounds. Then for a biological display, the chromatograms were divided into zones according to a preliminarily prepared scheme. Chromatogram and control zones equal in area and were placed into Petri dishes. Then they were moistened with distilled water and 10 cress germinating seeds were planted per each dish. After 24 hours the increase in cress roots was measured. Each test was repeated at least twice.

As per data, the high allelopathic activity of living Korean chrysanthemum was due to new phenolcarboxylic acids (chlorogenic) entering into their composition and mixture of flavonoids (quercitin and gossipetin, phenolic glycosides) (Table 39). Flavonoids also occured in plant residues as inhibitors (gossipetin, quercitin, isoquercitin, phenolic glycosides). Chlorogenic acid was present in very small amounts that it is unable to exert an inhibitory effect (Table 40). In Chinese aster, quercitin and phenol glycoside enter into the composition of kolines with phenolcarboxylic acids (Table 41). Thus, the high allelopathic activity of given plant species wa owing to complex of phenolic compounds. A comparison of the qualitative composition of kolines of aster and chrysanthemum showed a certain similarity between them. Both living plants and their residues contain quercitin and it acted as growth inhibitor. While the allelopathic activity of chrysanthemum was provided by gossipetin (8-oxi-derivative of quercitin (equally with other compounds), but toxicity of above-ground parts of living plants (aster and chrysanthemum) was provided by phenolcarboxylic acids.

Kolines did not remained in the soil in unchanged state, hence, allelopathic role of these or other compounds can be estimated only when their further fate was explained. Hence, we studied the influence of soil upon the activity of kolines of Chinese and Korean chrysanthemum.

The following methods were applied to conduct trials. Aqueous and ethanol extracts from the residues of Chinese aster and Korean chrysanthemum were passed through a soil layer (20 cm) taken from under the same cultures. The time of the extract contact with soil was 2 hours. The initial extracts after their contact with soil were also vapourized, the remains were dissolved in 70% ethanol and chromatographed in two solvent systems: n-butanol–acetic acid–water 3:2:95 and 40:12:28. In all the treatments, the soil has a significant influence on the qualitative composition of allelopathic substances contained in plant residues. In particular, as a result of a short-time contact of water extracts from the residues of dwarf queen red aster with soil, practically all the substances were absorbed. Soil adsorption of a substance with Rf 0.66–0.82 determines disappearance of a corresponding inhibitor. At the same time when extract was passed through soil, new substances appeared in it, part of which was active (substance with Rf 0.02–0.17; Fig. 8, Table 42). Less accurate data were obtained from the extract chromatography in the other system of solvents (n-butanol–acetic acid–water 3:2:95).

Table 39. Physico-chemical properties of inhibitors isolated from fresh aboveground parts of chrysanthemum W. (extracted with 70% ethanol), 1975

Compound No	Rf value	Colour in visible light		Fluorescence in ultraviolet light		Colour reaction				Presumable compound
		Visible light	Visible light + NH_3	Ultraviolet	Ultraviolet + NH_3	DSA	$FeCl_3$	$AlCl_3$ Visible light	$AlCl_3$ Ultra-violet	
Butanol-acetic acid-water (40:12:28)										
1	0.33-0.45	Light yellow	Yellow	Yellow	Brightly yellow	Yellow	Green	Yellow	Yellow	Quercetin
2	0.47-0.57	Light yellow	Yellow	Yellow and grey	Brightly yellow	Yellow	Grey and green	Yellow	Yellow	Gossipetin
3	0.86-1.00	Light brown	Light brown	Yellow and brown	Yellow and brown	Pink	Green and brown	-	-	Phenolic glicoside
Butanol-acetic acid-water(3:2:95)										
1	0.00-0.10	Light yellow	Yellow	Yellow	Brightly yellow	Yellow	Green	Yellow	Yellow	Flavonoid mixture
2	0.45-0.56	-	-	Blue and green	Brightly blue green	Yellow and pink	-	-	-	Chlorogenic acid
3	0.73-0.90	Light brown	Light brown	Violet	Violet	-	-	-	-	Phenolic glicoside

Table 40. Physico-chemical properties of inhibitors isolated from residues of chrysanthemum 'Leto'

Comp-ound No	Rf Value	Colour in visible light		Fluorescence in ultraviolet light		Colour reaction				Presumable compound
		Visible light	Visible light + NH_3	Ultraviolet	Ultraviolet + NH_3	DSA	$FeCl_3$	$AlCl_3$ Visible light	$AlCl_3$ Ultra-violet	
Butanol-acetic acid-water) (40:12:28)										
1	0.39-0.47	Light yellow	Dark yellow	Yellow	Brightly yellow	Yellow	Green and grey	Yellow	Yellow	Quercetin
2	0.47-0.54	Light yellow	Yellow	Dark	Yellow and green	Yellow	Green	Yellow	Yellow	Isoquercetin
3	0.54-0.63	Light yellow	Yellow	Grey and yellow	Yellow	Yellow	Grey and green	Yellow	Yellow	Gossipetin
15% acetic acid										
1	0.18-0.23	Light yellow	Yellow	Grey and yellow	Brightly yellow	Yellow	Grey and green	Yellow	Yellow	Gossipetin
2	0.80-0.90	Light yellow	Yellow	Brightly yellow	Yellow	Pink	Green	-	-	Phenolic glycoside
Butanol-acetic acid-water (3:2:95)										
3	0.75-0.85	Light yellow	Yellow	Yellow	More yellow	Pink	Green	-	-	Phenolic glicoside

Table 41. Physico-chemical properties of growth inhibitors isolated from residues of China aster

No	Rf	Colour in visible light		Fluorescence in ultraviolet light			Colour reaction			Solubility in		Presumable compound
		Visible light	Visible light + NH_3	Ultraviolet	Ultraviolet + NH_3	DSA	$FeCl_3$	$AlCl_3$		ethanol	water	
								Visible light	Ultra-violet			
1	0.00-0.05	Yellow	More yellow	Yellow	Brightly yellow	Yellow	Green and grey Braun	Yellow	Yellow and green	Good	Weak	Quercetin
2	0.65-0.85	Weakly brown	Brown and yellow	Weakly yellow	Yellow	Yellow and orange		-	-	Good	Good	Glyciside (aglycones: apigenin? campferol?)

Table 42. Results of chromatographic separation of water extracts from residues of red China aster from dwarf queen group

	Initial extract			After interaction with soil			
		Fluorescence in ultraviolet light				Fluorescence in ultraviolet light	
No	Rf Value	Ultraviolet	Ultraviolet + NH_3	No	Rf Value	Ultraviolet	Ultraviolet + NH_3
Butanol-acetic acid-water (40:12:28)							
1	0.00-0.10	Violet	Violet	-	-	-	-
2	0.10-0.26	Dark	Yellow and green	-	-	-	-
-	-	-	-	1	0.02-0.17	Brightly yellow	Without change
3	0.26-0.43	Yellow and grey	Yellow and green	-	-	-	-
-	-	-	-	2	0.17-0.30	Bright	Without change
-	-	-	-	3	0.30-0.37	Blue	Without change
-	-	-	-	4	0.37-0.42	Bright	Without change
4	0.43-0.63	Yellow	More yellow	-	-	-	-
5	0.66-0.82	Blue	Green and blue	-	-	-	-
6	0.82-0.98	Dark	Dark	-	-	-	-
				5	0.90-1.00	Violet	Without change
Butanol-acetic acid-water (3:2:95)							
-	-	-		1	0.02-0.08	Green and blue	More green and blue
1	0.10-0.22	Blue	More blue	-	-	-	-
2	0.24-0.32	Blue	More blue	-	-	-	-
3	0.33-0.39	Dark	Dark	-	-	-	-
4	0.45-0.53	Brightly yellow& green	Without change	-	-	-	-
-	-	-	-	2	0.50-0.56	Azure and blue	Without change
5	0.55-0.64	Yellow	More yellow	3	0.65-0.69	Weakly yellow	Yellow
6	0.64-0.80	Dark	Without changes	4	0.70-0.80	Dark	Without changes

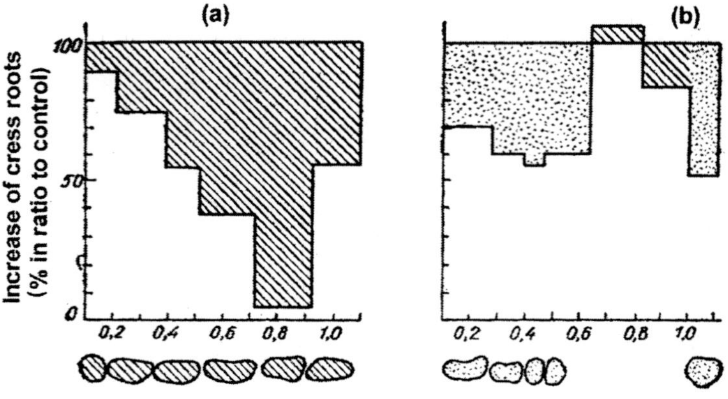

Fig. 8. Histogram of biological activity of water extract from dwarf queen aster residues (butanol-acetic acid-water 40:12:28), 1977.

Kolines of Korean chrysanthemum also undergo essential changes in soil (Fig. 9, Table 43). The soil quite actively absorbed the flavonols 1–4 and phenolcarboxylic acids (compounds 5–7). Absorption of substances with Rf 0.10–0.25, 0.44–0.52 and 0.52–0.63 led to disappearance of inhibitors; partial absorption of substance with Rf 0.83–0.98 caused a significant decrease in inhibitory activity of a given koline. The display of new substances with Rf 0.03–0.20 and 0.25–0.40 means the appearance of earlier absent substances. The chromatography of both types of extracts in the system n-butanol–acetic acid–water 3:2:95 was to be less successful.

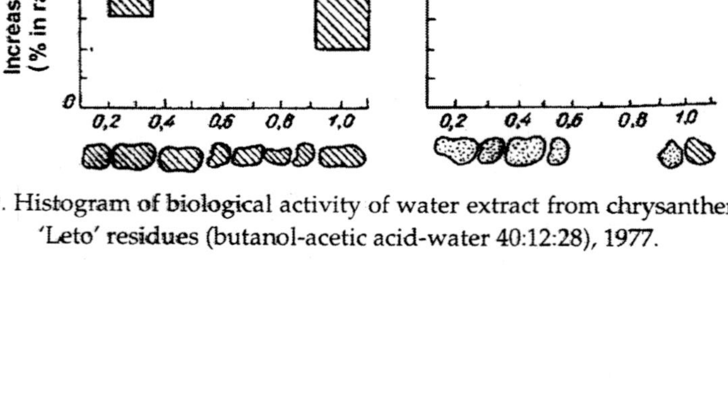

Fig. 9. Histogram of biological activity of water extract from chrysanthemum 'Leto' residues (butanol-acetic acid-water 40:12:28), 1977.

Table 43. Results of chromatography separation of water extracts from residues of chrysanthemum 'Leto'

Initial extract				After interaction with soil			
		Fluorescence in ultraviolet light				Fluorescence in ultraviolet light	
No	Rf	Ultraviolet	Ultraviolet + NH_3	No	Rf	Ultraviolet	Ultraviolet + NH_3
Butanol-acetic acid-water (40:12:28)							
1	0.00-0.10	Violet	Without change	-	-	-	-
-	-	-	-	1	0.03-0.20	Sand yellow	Without change
2	0.10-0.25	Dark	Yellow and green	-	-	-	-
3	0.26-0.44	Grey and yellow	Yellow and green	-	-	-	-
4	0.22-0.52	Yellow	More yellow	-	-	-	-
5	0.52-0.63	Blue	Green and blue	-	-	-	-
-	-	-	-	3	0.25-0.40	Blue	More blue
-	-	-	-	4	0.42-0.47	Dark	Yellow and green
6	0.63-0.73	Blue	Blue	-	-	-	-
7	0.73-0.83	Green and blue	Without change	-	-	-	-
-	-	-	-	5	0.80-0.88	Violet	Without changes
8	0.83-0.98	Violet	Azure	6	0.90-1.00	Violet	Azure
Butanol-acetic acid-water (3:2:95)							
1	0.00-0.05	Violet	Without change	-	-	-	-
2	0.18-0.24	Weakly blue	Blue	-	-	-	-
3	0.64-0.70	Blue	More blue	1	0.65-0.71	Blue	More blue
4	0.70-0.80	Yellow	Yellow and green	2	0.71-0.85	Yellow	More blue
5	0.80-0.97	Dark	Without change	-	-	-	-

Though water is natural extraction solvent, we conducted an additional series of trials with the use of ethanol extracts from plant residues. Because it has been proved in our investigations, that the best way of koline excretion from plant material was the extraction of physiologically active substances by 70% ethanol. As indicated by investigations, soil adsorbed flavonoids from koline complex of the studied objects (quercetin, isoquercetin, gossipetin). In this manner, we found that the fate of kolines of flower plants in soil was different. Part of them (basically with high Rf values in acid system) were not absorbed by soil, under natural conditions. They were leached into deeper layers, i.e. similar compounds have a comparatively little significance to trigger off allelopathic effects. From the allelopathic point of view, more interesting was those kolines which were absorbed by soil colloids (quercitin, isoquercitin and gossipetin), because the accumulation of toxic products in upper layers of soil was the main condition of allelopathic soil sickness owing to excretion of physiologically active substances, formation of toxins during the decomposition of plant residues and as a result of microflora activity, new and new allelopathic particles enter the soil. Due to exchange process, these substances were adsorbed in soil and displace the earlier adsorbed kolines. Hence, allelopathic substances were constantly present in soil both in, free and adsorbed fraction.

Determination of kolines in plants (or in residues) were their identification was not enough to determine their allelopathic effect. Our investigations have shown that there are allelopathically active substances in soil, which were absent in plant itself. The determination of quantity of kolines in soil and their chemical composition was the primary task of any question concerning allelopathy. Using a new method of displacement we have revealed the isoquercitin (Rf 0.41–0.50) in the sick soil (from under the chrysanthemum cv. 'Leto'), which entered into the complex of chrysanthemum allelopathic substances, many phenolic substances are present in living plants and in their residues. Besides the substances absent in plant material were also found in the soil.

We found that large number of phenolic substances accumulate in the soil under aster plants. We must determine the quercitin in sick soil, because quercetin along with other substances provides a high allelopathic activity of aster.

2.2. Nature of Kolines

Our investigations separated the following groups of active allelopathic compounds.

Group I : These compounds cause toxicity only in living plants and after entering the soil, they gradually break down e.g. chlorogenic acid, which provides a high allelopathic activity of above-ground parts of Korean chrysanthemum. Plant residues also contained the phenolcarboxylic acid, but in very low concentrations to exert an inhibitory effect. Chlorogenic acid was not found in the soil.

Group II: These substances of phenol glycoside type enters into complex of allelopathically active substances of living plants and determine the toxicity of plant residues of aster and chrysanthemum. Phenolic glycosides were not found in the soil.

Group III : These are flavonol type compounds. We have found that gossipetin, isoquercitin and quercitin are present in living Korean chrysanthemum. The gossipetin and isoquercitin accumulate in soil. Quercitin occurs as allelopathic substance in Chinese aster plant and in the soil under it. Besides, similar substances are more stable in soil. According to our investigations, they did not penetrate deep into the soil, but saturate it and retains their allelopathic activity. The flavonols in course of passing from bound to a free state can be adsorbed on other compounds (root excretions and decomposition products of koline precursors or microflora metabolites) by exchange adsorption. The kolines accumulating in soil, under chrysanthemum are isoquercitin and gossipetin and under asters is quercitin.

Group IV: These kolines consists of substances, which are absent in plant itself or in plant residues, but accumulate in toxic concentrations in soil and play certain role in the occurrence of allelopathic soil sickness. These are products of decomposation of those compounds which comes into soil with the plant material or appear as a result of intensive activity of soil microflora).

In this way, in repeated (monoculture) flower growing in the same place, allelopathic soil sickness occurs due to accumulation of physiologically active substances of plant and other origin under crop plants.

Chapter 5

SOIL TOXINS : CHEMICAL NATURE

A.M. Grodzinsky

Department of Allelopathy,
Central Botanical Garden, 1, Timirvzskaya Street, Kiev, Ukraine

> *"Any art doesn't cure the soil which has become barren...Its mellowness doesn't keep introduced humidity, its dryness prevents plants from using manure and inherent bitterness interferes with nutrition of seeds and plants."* Albert the Great, 1256. Treatise on plants. On treating field to make it fertile. In: Agriculture in Journals of the west Middle Ages, 1936.

Chemical interactions of plants in space and time are carried out in different ways with participation of not only higher plants, but also diversified heterotrophic organisms, soil and many other factors (Grodzinsky, 1972). The basis of ecological mechanisms of allelopathy kolines are accumulation of physiologically active products in soil.

Rice (1974) had divided inhibitors involved in chemical interaction into following 15 classes:

A. Simple water soluble organic acids, alcohols with unbranched chain, aliphatic aldehydes and ketones.
B. Simple unsaturated lactones.
C. Fatty acids with a long chain.
D. Benzoquinones, antraquinone and complex quinones.
E. Terpenoids and steroids.
F. Simple phenols, benzoic acid and their derivatives.
G. Cinnamic acid and its derivatives.
H. Coumarines.
I. Flavonoids.
J. Tannins (hydrolized and condensed).
K. Amino acids and polypeptides.
L. Alkaloids and cyanohydrines.
M. Sulphides and glycosides in mustard oils.
N. Purines and nucleosides.
O. Miscellaneous (substances present in two or three groups simultaneously, and substances not entered into any of these classes e.g., ethylene).

All these substances can probably be encountered in soil sickness. However, it is difficult to ascertain the chemical nature of substances causing soil sickness. It is a complex and frequently a questionable matter, because there are enormous difficulties in their extraction from soil, in their cleaning and identification and at least in proving that just these identified substances determine the allelopathic effect.

There are numerous observations and direct and indirect proofs that allelopathy of plants in natural communities helps in accumulation of kolines at high concentrations in soil. The reviews of French geobotanists, given in the book by Grümmer (1957), proved the accumulation of toxic substances in soil under the *Brachipodium pinnatum* P.B. turf, under developing flower-beds of *Hieracium umbellatum* L., under *Rosmarinus officinalis* L. bushes, etc. We have found that under some extremely active allelopathic steppe plants, such as *Salvia austriaca* L., *Crambe tataria* Sebeök., *Adonis wolgensis* Stev., *Salvia nutans* L. and other, active kolines accumulate so abundantly, that these plants can grow only alone and their place growing should be changed for each generation (Grodzinsky, 1969).

Rice (1974) proved the accumulation of chlorogenic acid in the soil under wild sunflower in fallow lands and under some scrub species and the accumulation of tannic acid – under *Rhus glabra, Rh. copallina*, etc. These substances not only displayed allelopathic effect, but also depressed the nitrogen fixing and nitrifying organisms, which led to serious nitrogen deficiency. Everything taken together, created favourable conditions for succession of new weeds and steppe species.

There is available information in the literature about the accumulation of physiologically active substances in soil under field crops. Mishustin (1961) and Naumova (1959) revealed the accumulation of saponins in soil under alfalfa. Stepanova (1971) confirmed this observation and showed the mechanism of toxin action on flax. Substances contained in flax residues inhibited its growth and reduced productivity (Grümmer, 1970). Unknown inhibitors, which accumulated under sugar beet, displayed their effect on maize crop next year (Dzubenko and Boyko, 1972).

Unfortunately, it is very difficult to prove, that in such cases accumulation of toxins is the direct reason of soil sickness. Initial plant poisoning takes place very gradually and it is difficult to notice it. Weakened plants lose resistance to cold or drought, their immunity to pathogens and pests decreases, their competitiveness in relation to weeds declines. Diseases, pests and weeds appear in mass. They are well observed, hence, they are considered as the main cause of soil sickness, though they are only intermediate links in the entire chain of failures. Disease and pest suppression as well as weed control are a little success, because the initial cause – toxicity of soil still remains and the entire rotation of factors unfavourable to plants is started again. Koline accumulation in soil takes place also as a result of changed ecological conditions. Soil sickness can also be induced by uneven development of saprophytic microflora in soil.

In our laboratory, Yourtshak (1971) reported that microflora decomposing residues of ploughed alfalfa as a green fertilizer, also releases many toxins from macerated cells and produces its own phytotoxic metabolites and because of that the soil inhibited the plant growth for at least two weeks after reversible ploughing.

One of the most common mechanisms of allelopathy is creating the defence zone around the donor-plant, or if many plants grow together and their zones merge together, there occurs allelopathic field. The potential of allelopathic field fluctuates in relation to organic product supply to the environment, fluctuations of temperature, humidity and other factors, which determine accumulation and degradation of kolines. In forest association, allelopathic field is composed of substances leached by sediments from living leaves, root excretions and mainly from the compounds of forest litter (Tukey, 1966, 1969). The presence of such forest litter determines the absence of grass cover and a strong depression of juvenile plants under close red beech, spruce and pine plantations, under separately standing trees of walnut, plane-tree, eucalyptus and other trees.

Substances leached by dew- and fog-drops from shrub leaves in mountains of California determine a complete depression of grassy vegetation as well as shrub seedlings of *Adenostoma fasciculatum* and *Arctostaphylos glandulosa* (Müller *et al.*, 1968). Cutting down trees and shrubs or burning them leads to a situation, in which the release of kolines is stopped and lush development of grasses begins. Shoots of trees and shrubs "sitting upright" plants until that time in a juvenile stage begin to grow. If fallen needles in the pine forest are removed periodically, the soil surface will be covered with grasses, the seeds of which are present abundantly in the soil, but do not germinate in the presence of litter. Chemical nature of substances, which determine the suppression of grass and juvenile growth and keep seeds in dormancy, has been determined in a number many cases (Table 44).

With the fall of wood plants, numerous phenolic compounds are added into the environment. Since phenols strongly depress the development of the majority of bacteria and to a small degree influence the fungus metabolism. Thus decomposition of litter in forest is done mainly by fungi, which produce many organic acids. This determines podzol type of soil formation. Besides, powerful endotrophic or ectotrophic mycorrhizas develop on the tree roots and they suppress the bacterial flora by antibiotic exometabolites. Phenols are decomposed by fungi, but not all and not at once. Some quantity of mobile forms of plants depressing the appearance of shoots, growth and development of woody plants and grassy species are constantly present in the environment. This mechanism of plant interaction occurs in areas of sufficient precipitation, where easily

Table 44. Chemical nature of compounds creating allelopathic defence zone under tree plants

Plant species	Mode of release	Compound	Biological activity	Author
Adenostoma asciculatum	Leachates from leaves	Phenolics	Growth inhibition	Mc Person, Müller (1969)
—do—	Leaf fall, root residues	Esculin, esculetin, catechins	Growth inhibition	Oleksyevitsh (1971)
Aesculus hippocastanum	Leachates from leaves, fall, soil under tree crown	Arbutin, phenolics	Inhibition of seed and bulb germination	Müller, Hanawalt, Mc Person (1968); Chou and Müller (1972)
Backhousia angustifolia	Leaf fall	Volatile triterpenes, angustion	Growth inhibition	Cannon *et al.*, (1962)
Berberis vulgaris	Root residues	Berberin, berberitin	Growth reduction, mitosis disturbance Reduction of vegetation creation of bare zones	Oleksyevitsh (1971)
Eucalyptus vulgaris	Fall, soil, volatiles	10 phenolics including coffeic, p-coumaric, chlorogenic and gallic acids	Reduction of vegetation creation of bare zones	Moral del, Müller (1970)
Eucalyptus globulus	Dew, fall, soil	Chlorogenic, p-coumaric acids, volatile terpenes	Reduction of vegetation creation of bare zones	Moral del, Müller (1969)
Juglans nigra, J. regia L.	Fallen leaves, leachates	Juglone	Oppression of different plants, harmful effect on flies and other animals	Bode (1958)
Malus domestica	Root residues, fallen leaves, cut twigs	Phloridzin, phloretin, hyperin, floroglucin, floretinic, p-hydroxy-benzoic, chlorogenic and coffeic acids	Oppression of apple and other plants, changes of soil microflora composition	Börner (1959, 1960); Moroz (1968)
Parthenium argentatum	Root exudates	Cinnamic acid	Oppression toward plant of the same species and other species	Bonner (1946b, 1950)
Persica vulgaris	Root residues, fallen leaves	Catechins, astragalin, izoquercetin, campferol, chlorogenic and coffeic acids, amygdalin	Soil sickness during and after replantation	Patrick (1955); Moroz (1968)

water soluble phenolic substances, are periodically leached from the tree crown, bark and from the roots - by rains, dew, melted snow and fog etc.

A completely different ecological mechanism has appeared in the areas with insufficient precipitation and partial overheating of plants. Müller (1970) and his co-workers showed that, in the arid mediterranean climate of California, where there was practically no rains from May to November, essential oils evaporated by shrubs (*Salvia leucophylla, S. mellifera, S. apiana)* are absorbed by dry soil and preserved in it, until the beginning of the rainy season in November, when a vigorous development of annual grasses begins in the prairie. The presence of terpenes in soil determines its extremely toxic properties and there was a 1 m bare strip around shrub undergrowth, where nothing is growing. Approximately the same mechanism of volatile substances has been found for California wormwood and for eucalyptus (*E. camaldulensis* – del Moral and Müller, 1970; *E. microtheca* – Al-Mousawi and Al-Naib, 1976). The identified substances are presented in Table 45.

Table 45. Chemical nature of volatiles accumulated by soil in surrounding aromatic plants

Species	Compound	Biological activity	Author
Artemisia absinthium	β-caryophylen, bisabolen, hamazulen	Growth inhibition of neighbouring plants	Schwar (1962)
Eucalyptus camaldulensis	Cineol, α-felandren, α-pinen, β-pinen	Common growth inhibition, inhibition of seed germination	Moral del, Müller (1970)
Eucalyptus microtheca	α-pinen, β-pinen, camphen, cineol	Growth inhibition of neighbouring plants, inhibition of seed germination	Al-Mousawi, Al-Naib (1976)
Salvia leucophylla, S. apiana, S. mellifera	Cineol, α-pinen, β-pinen, camphen camphora, dipenten	Growth inhibition of roots, reduction of mitochondrial activity	Müller and Müller (1964); Müller *et al.* (1968)

With continuous soil watering or during long lasting rains, this allelopathic mechanism ceases to act, because moist soil does not absorb essential oils. However, even under conditions of normal soil moisture, the soil periodically gets dry, and just in these periods, significant amounts of lipophilic physiologically active compounds can accumulate in the soil. Sometimes, these compounds are not especially volatile and at the next moistening of soil they remain in the place and the soil become "fatty" in the literal meaning of this word. Even its water retaining property is changed (Rice, 1974; see also chapter 4). Schreiner and Reed (1908) introduced a new concept – "agricer", or soil wax, which can be removed by organic solvents from the exhausted soil.

Svishtshhouk and co-workers (1977) identified 33 fatty acids in the apple-sick field in the lipid soil fraction (0.14% of the soil mass), whereas, in the neighbouring field under centuries-old oak the total lipid content was 2.5-fold lower and appeared to be insufficient to identify fatty acids on gas chromatography.

M.A. Pantshuk from our lab, revealed the presence of $C_{12}H_{18}$ unusually active triterpene, volatile and lipophilic in the soil under wheatgrass (*Agropyron glaucum*) and under some wheatgrass hybrids and selections. Probably the mechanism of volatile lipophilic substance absorption was quite common. We have observed that in humid years Austrian sage (*Salvia austriaca*) in Homutovska steppe did not show suppression effect on the shoots of steppe plants including its own ones, though it usually distinguishes by a high allelopathic activity. Probably this can also explain the fact, that a high toxicity of soil observed in nature in rosemary-heath associations in the south of France was not supported by Négre (1968) in the vegetation experiment, where plants were regularly watered.

Rice (1974) described a special mechanism of plant interaction during overgrowing of long-fallows in Oklahoma and Kansas. Many large weeds occupying field in the first fallow succession, excrete different phenolic inhibitors (Table 46), which depress and force out the less stable species and also cause autotoxicity e.g. sunflower (*Helianthus annuus*). These inhibitors also suppress the nitrogen fixation and nitrification by symbiotic and free-living micro-organisms, whereas, denitrification remains unchanged. Thus nitrogen supplies in soil are rapidly exhausted and the dominating plants in the second stage of succession will be grasses with minimal requirements for nitrogen. Weeds produce noticeable amounts of kolines into the soil, which influence the crop plant, either immediately, or in the next year, i.e. they cause soil sickness (Table 46).

Table 46. Inhibitors of growth and nitrogen fixation produced by weeds

Weed species	Plant part	Compound	Author
Digitaria sanguinalis	Whole plant	Chlorogenic, izochlorogenic acids	Parenti and Rice (1969)
	Living whole plant	Chlorogenic, isochlorogenic acids, sulfosalicylic acid Gallic and tannic acids	Rice (1965)
Euphorbia corollata Sorghum halepense	Leaves, rhizomes	Chlorogenic, p-coumaric acids, p-hydroxybenzaldehyde	Abdul-Wahab, Rice (1967)

Thus, in this mode of allelopathic interaction, the main role is played by phenolic compounds.

Studies of Plhak (1972) and Kovaltshouk (1973) showed the presence of ethylene in volatile root excretions and in the soil. It may be suggested, that biogenic ethylene participates in some forms of soil sickness. Ethylene is highly active growth substance and influences various processes, causing infrequently a premature leaf fall and different morphological deformities. Under the influences of ethylene on pea seedlings, the activity of peroxidase intensifies and the process of lignification in pea seedlings becomes slower (Kovaltshouk and Bukolova, 1975). Since peroxidase is the main component of IAA-oxidase, hence, the authors suggest that the inhibiting ethylene influence the growth of pea seedlings due to oxidizing break down of IAA intensified peroxidase activity. After ethylene action, the activity of peroxidase decreases and becomes normal and the growth processes restarts.

The clover sickness was determined by accumulation of phenolic acids, isoflavonoids and similar substances in soil (Chang *et al.*, 1969; Tamura, 1969). Accumulation of saponins in soil under monoculture of alfalfa has negative influence on the seed germination of cotton sown in stubble mulch, whereas, cereal crops do not suffer from saponins (Mishustin and Naumova, 1975).

In our laboratory L.P. Stepanova has confirmed the saponins accumulation under alfalfa and has showed that other substances of phenolic origin also participate in soil sickness. The different alfalfa cultivars were identical in their ability to exhaust the soil. It indicates the possibility of selecting the cultivars, which will cause minimal deterioration of soil fertility.

Post-harvest residues ploughed into the soil are a very important source of kolines and serve as a food for microflora. We analysed 200 samples of various post-harvest residues collected from the fields of Kiev district (Grodzinsky *et al.*, 1962, 1963). It appeared that the majority of them contained substances inhibiting seed germination as well as growth of shoots and roots. These remains are a source of physiologically active substances toxic to crop plants. In the vegetation experiment, negative effect of applied post-harvest residues on plants disappeared after 2-3 weeks, their growth improved and at harvest gave yields equal to control treatment. Apparently, due to decomposition of inhibitors released from the post-harvest residues, their inhibitory effects changed to stimulatory and everything became more equal. However, under field conditions, the delay in growth caused by the inhibitors from post-harvest residues may lead to yield reduction. Various post harvesting residues (stubble and straw of leguminous plants, vegetables, potato tops, cut branches of vines, green forage, hay etc. should not be considered as manure, because they immobise the nutrient substances till their complete decomposition.

McCalla and his co-workers have found that stems and other residues form a layer on the soil surface, which during the winters drastically depresses the growth of maize, wheat, alfalfa and other plants. Because the fungus *Penicillium urticae* Bainer, which produced patulin have strong inhibitory effect on higher plants, developed on harvest residues (Guenzi and McCalla, 1966 a,b; McCalla, 1971). Patulin was found in ethyl acetate extracts of straw decayed at concentrations of 25–50 μg/g in the soil. In laboratory experiments conducted in the unsterilized soil, patulin decomposed very rapidly. The period of half-decay was 24 hrs. Maintenance of high concentrations of patulin was determined by stable synthesis and its release in the surroundings.

1. KOLINES

1.1. Amino acids

A large number of amino acids are released into the soil environment during decomposition of plant residues containing proteins. Amino acids are also present in root and seed excretions. They participate in the physiological activity of substances from the soil extract on the plants. Separate constitutional amino acids under sterile conditions inhibits the growth of higher plants (Ratner *et al.*, 1963), however, in combined application of two and more amino acids, no toxicity occurred. On the contrary, valine, d-phenylalanine, asparagine and glutamic acids, dl-lysine, d-arginine, dl-serine, l-tyrosine, dl-proline, l-hydroxyproline, l-histidine and cysteine proved stimulatory (Dietz, 1959). However, some amino acids e.g. methionine and particularly unique and unnatural amino acids and compounds similar to them, may be toxic to plants (Aldag, 1966). Apparently, a biochemical mechanism of toxicity of those unusual amino acids consists in that they are included into a normal amino acid metabolism of plant cell and "close" it, because they block the enzymes and have no further changes. Some cyclic polypeptides synthesized by microscopic fungi, for example, malphormine has great physiological activity.

Winter and Street (1963) reported strong growth inhibitor, separated from an old cultivated site where tomatoes were grown. This substance was cyclic polypeptide with vanillin residue in molecule.

Samsevich and Borisova (1963) showed in many laboratory experiments that volatile soil excretions depressed winter wheat shoots and development of *Azotobacter*. This activity increased after composting of soil with high moisture and temperature and especially after the application of organic substances (humus, sucrose), i.e. it was determined by microflora activity. Toxicity was caused by ammonia, fatty acid and other substances. These substances, regulate the quantitative and qualitative composition of soil microflora.

However, only control soil composting, i.e. its maintenance under moistened conditions at thermoregulation, which activates soil microflora, increases its own toxicity even without organic substances (Norstadt and McCalla, 1968).

Agricultural activity of farmers frequently leads to peculiar soil sickness. Classical examples are, soil sickness under repeated monocultures of flax, beet, pea, millet, clover and other plants. Strong toxicity appears at cutting down of trees and shrubs or at moving grassy plants, when significant amounts of root residues remained in the soil. Thus, about 6–10 t/ha of roots remain in the soil after cutting down old trees of apple orchards. The roots contain 14–30% of phloridzin i.e. 1–2 t/ha of that highly active phenolic compound will remain in the root-zone of soil and will be gradually released from the dead tissues into the surroundings. Four weeks after, phloridzin is completely decomposed due to microbial activity.

In composting under laboratory conditions, patulin (Norstadt and McCalla, 1968) and trans-cinnamic acid (Bonner, 1950) disappear very quickly, which gave a reason to some authors to doubt about the evidence of allelopathic activity of some compounds (Bonner, 1950; Müller, 1958; Grümmer, 1957). However, under natural conditions, these processes are not so quick and besides that, toxins are constantly replenished either from slowly decomposing residues of apple roots and peach (Moroz, 1968), from harvest residues infected by fungal mycelia (*Penicillium urticae*) or from root excretions of living plants of guayule (*Parthenium argentatum*). For this reason, the toxins are present in the exhausted soil for quite a long time (from one to several seasons) causing damage to agriculture.

1.2. Phenolic acids

The phenolic substances mainly acids, are mediators of many allelopathic mechanisms, but these were least investigated as possible factors of soil sickness. Phenolic acids and their derivatives are generally found in the soil extracts. Walters (1917) reported about the presence of p-hydroxybenzoic and benzoic acids in Florida soils exhausted under citrus plants. Wang and co-workers (1967) have found p-coumaric, ferulic, p-hydroxybenzoic, vanillic and syringic acids and number of unidentified phenolic acids in some soils of Taiwan under sugar-cane revealing symptoms of soil sickness. Whitehead (1964) also found p-hydroxybenzoic, p-coumaric, vanillin and ferulic acids in soil.

Wang and co-workers (1967) determined the content of phenolic acids in soils of Taiwan where sugar-cane was grown without organic fertilizers (Tainan in Taiwan), as well as under tobacco, batata, pineapple and banana. The total content of phenolic acids in the arable layer under sugar-cane ranged from 5.28 (red soil) to 45-47 μg/kg of soil (yellow and alluvial soils). In deeper layers their content decreased, especially in sandy soil. It is interesting, that under cane rows, the content of phenolic acids was higher than that in the interrows (Table 47).

Table 47. Content of phenolic acids (μ moles/kg) in soils from under rows of sugar cane and in interrows (by Wang *et al.*, 1967)

Soil and place	Depth (cm)	Acids present			Total
		p-Hydroxy-benzoic	Vanillic	p-Coumaric	
Alluvial No 6					
Under rows	0-25	17.98	12.99	4.66	35.63
	25-50	1.83	0	1.60	2.43
Interrows	0-25	0.83	0.17	0	1.00
	25-50	0.83	0.10	0	0.83
Alluvial No 7					
Under rows	0-25	1.03	10.32	0	1.03
	25-50	0.53	0	1.57	2.10
Interrows	0-25	1.03	0.17	1.98	2.10
	25-50	1.70	0.10	3.0	4.80
Alluvial No 8					
Under rows	0-25	10.66	10.32	16.98	37.96
	25-50	0.40	0	1.03	1.43
Interrows	0-25	1.32	0.17	1.97	3.46
	25-50	0.60	0	0.40	1.00
Alluvial No 9					
Under rows	0-25	0.83	0.67	19.98	18.48
	25-50	0.63	0.27	1.03	1.93
Interrows	0-25	1.00	0	12.99	13.99
	25-50	0.40	0.93	0	1.33

This regularity is observed also under tobacco and is absent under other crops (Table 48). The content of p-coumaric acid significantly exceeded that of vanillic and p-hydroxybenzoic acids, ferulic acid content was still lower, whereas, syringic acid occurred in trace amounts. Taking into account a strong inhibitory activity of phenolic acids on plants, their similar concentration in soil should be toxic. It is obvious that the authors also extracted bound forms of phenolic acids, which are not practically dangerous.

Table 48. Content of phenolic acids (μmoles per kg) in ploughed layer under different cultures (Wang *et al.*, 1967)

Crop	Acids			Total
	p-Hydroxy-benzoic	Vanillic	p-Coumaric	
Tobacco				
Under rows	2.86	3.00	22.64	28.50
Interrows	3.66	3.00	10.99	17.65
Batata				
Under rows	0.40	0.27	0	0.67
Interrows	3.66	2.03	1.97	7.66
Pineapple				
Under rows	2.86	1.37	28.31	32.54
Interrows	3.66	1.60	44.62	49.88
Banana				
Under rows	0.83	0.80	0	1.63
Interrows	1.00	0.33	0.40	1.73

Hennequin and Juste (1967) reported that the content of three phenolic acids in the upper layers of podzol, brown, humus and carbonated soils (rendzina) ranged between the following limits: p-coumaric acid : 4.7-6.3 mg/kg, vanillic : 2.6-6.5 mg/kg, p-hydroxy-benzoic : 2.6-7.3 mg/kg. The deeper soil layers (40–80 cm) as rule contained less phenolic acids.

The application of high doses of organic fertilizers (300-350 t/ha) slightly increased the content of p-hydroxybenzoic acid, by 1.5-fold of vanillic and by several times more than that of p-coumaric acid. The authors noted that maize stems ploughed in soil are source of p-coumaric acid, which can be a reason of soil sickness under this crop culture observed in south-western France.

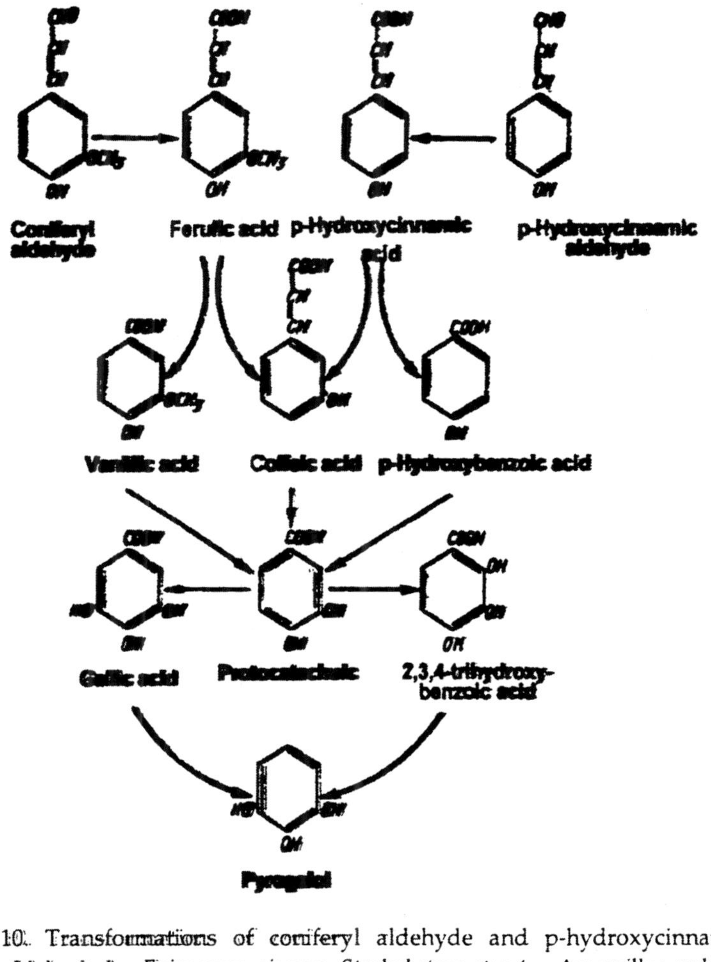

Fig. 10. Transformations of coniferyl aldehyde and p-hydroxycinnamic aldehyde by *Epicoccum nigrum*, *Stachybotrys atra* (or *Aspergillus sydowi*) (by Martin and Haider, 1971).

Phenolic acids can be formed during the microbial degradation of lignin. Figure 10 presents one of possible schemes of such transformations proposed by Martin and Haider (1971). Certainly, this scheme possibly reflects only a small part of transformations, because in the culture medium of the same fungi in glucose or in the residues of maize stems, numerous different phenols have been found depending on the source of carbohydrate nutrition, including.

Table 48A. Phenolic compounds released during the microbial decomposition of lignin

Phenolcarboxylic acids	**Toluols**	**Others**
cresylic	2,4-dihydroxytoluol	floroglucin
2,4-dihydroxybenzoic	3,5-dihydroxytoluol	5-methylpirogallol
2,5-dihydroxybenzoic	4-methyl-2,6-dihydroxytoluol	pirogallol
2,6-dihydroxybenzoic	2,3,5-trihydroxytoluol	resorcin
3,5-dihydroxy-3-methylbenzoic	2,4,5-trihydroxytoluol	
	2,4,6-trihydroxytoluol	
3,5-dihydroxybenzoic		
ferulic		
gallic		
6-methylsalicylic orselinic		
p-hydroxycinnamic		
protocatechuic syringic		
2,3,4-trihydroxybenzoic		
2,4,6-trohydroxybenzoic		
vanillic		

Besides, other unidentified phenolic compounds were also found.

Many simple phenols, which develop during the microbial decomposition of lignin or of constant humus or were synthesized by soil organisms, easily transform into active phenolic radicals and quinones due to the activity of phenoloxidases. These enzymes are produced by some basidiomycetes causing white rot. Phenoloxidase detaches hydrogen from hydroxyphenol structure. The radial formed in such way is stabilized either by linking with same radical or further transformation into quinone:

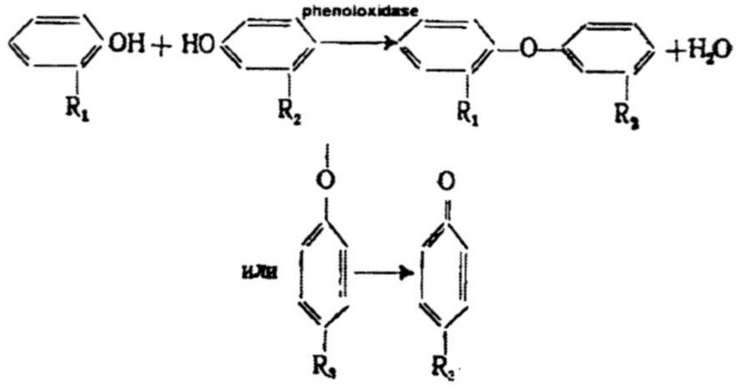

Quinones react with other phenolic compounds by a ring linking with amino acids and peptides producing polypeptides. Stom (1975) hypothesised that not only phenols but also quinones (formed during their oxidation are 100-fold more inhibitory to growth of higher and lower plants than phenols), are active in allelopathy. According to this hypothesis, phenolic compounds were inactive before oxidation. Depending on the presence of specific phenoloxi-dase, conditions of the environment and phenol state, quinones can be produced at different speed. This hypothesis drew attention of many scientists, because phenolic acids and other phenolic substances in concentrations existing in nature usually have no influence on test-objects and cause only a temporal inhibition of their growth. Therefore, a stable toxicity of soil can be determined by the presence of quinones. Products of incomplete oxidation formed during the complete or partial soil anaerobiosis are of no small importance, especially during flooding. There is a peculiar disease of rice called akiochi, which frequently occurs in Japan. At the shooting stage, the plants suddenly become dark, droop and die. This occurs in the fields containing large amounts of humus (peat) due tc application of abundant green or other organic manure (Takijima, 1964 a,b). Microflora fermenting the organic residues without oxygen develops under anaerobic conditions. Alcohols, hydroxycarboxylic acids, methane, hydrogen sulphide, ferrous oxide compounds and other harmfully acting substances are produced here. Production of lower aliphatic alcohols in the fermentation processes during the decay of plant residues, especially in oxygen deficiency is quite possible. Alcohols can also occur due to hydrolytic deamination of amino acids and aliphatic amines.

1.3. Alcohols

Studies of alcohols in soil have not been conducted. Wang and Chuang (1967) found no alcohols in four types of Taiwan soils taken from natural conditions. However, when composted at 1 g of dry leaf powder of sugar cane and 1.5 g of ammonium sulphate per 100 g of soil added 250 ml of distilled water, or 5 g of fresh leaves of *Crotalaria juncea* with the same amount of soil and water at 30°C, during the 8 weeks noticeable amounts of alcohols were found. Methanol, ethanol and n-propanol in the experiment with sugar cane and n-butanol were found with *Crotalaria*. The maximum was formed during the first weeks of composting. Methanol was predominant in the mixture in the beginning, then it gradually passed to higher alcohols, but by the end of the experiments they almost disappeared. n-Butanol reached its maximum after 4 weeks of composting. Maximum total amounts of alcohols in different soils composted with sugar cane ranged from 37.6 to 151.1 micromoles per 100 g of soil. With *Crotalaria juncea*, the respective amount was 1105.1-1659.7 micromoles per 100 g of soil mainly on the account of butanol.

The same soil composted under anaerobic conditions (water was added to soil to maintain 50% field capacity) also led to alcohol production. The maximum amounts for 4 types of alcohols with sugar cane leaves were : 21.7–78.9 μmole/100 g, for *Crotalaria juncea* – 58.2–256.3 μmole/100 g, i.e. less than that the soil flooding. However, in this case alcohols persisted for longer period in soil.

The authors tested the activity of methanol, ethanol, n-propane at concentrations of 4×10^{-4}- 4×10^{-2} M in the water culture on Hoagland solution on the shoot growth of sugar cane, maize and rice. It appeared that the released amounts of alcohols can be strongly toxic to plants. Unfortunately, very little data about the presence of alcohols in soil does not permit to evaluate the ecological importance of this group of substances.

1.4. Organic Acids

The amount of organic acids produced during the oxidation and metabolism of plants, animals and microorganisms in soil, as a rule, is not large, because they are easily metabolized by the majority of heterotrophic organisms. However, under partial anaerobic conditions, the application of large amounts of organic substances or other conditions temporily increases the organic acids to toxic levels.

Wang and his co-workers (1967) did not find organic acids in the control soil, but with composting of fresh leaves of *Crotalaria juncea* (5g per 100 g of soil ± 50% of field water capacity or adding 250 ml of water in soil) he found the following volatile acids: formic (0–0.29 μmole/100 g of soil), acetic (to 2.16), isobutyric, butyric, valeric, isovaleric – in trace amounts; unvolatile acids: oxalic (traces), vinous (tartaric acid), apple (malic) acids (to 6-73 μmole/100 g of soil). Monocarboxylic acids inhibited, while the hydroxylic and dicarboxylic acids stimulated the growth of sugar cane seedlings. Probably organic acids cannot occur independently as soil toxins, but their participation as synergists of more active phenolic acids is quite possible.

Root excretions of plants generally have an acidic nature. Large numbers of organic acids, which determined the process of podzol formation, are produced during fungal decomposition of forest fall litter. For this reason organic acids can play a special role in allelopathic soil sickness and attention should be paid to this fact in the future.

The gas chromatography have made it possible to identify volatile compounds of sulphur excreted by soil, mainly aliphatic sulphides and disulphides and other substances.

2. HUMUS

Soil humus is resistant to microbiological decomposition. In matured natural systems the amount of humus reaches an equilibrium state, i.e. the mass of created and mineralizing organic substance is balanced after a year. However, in unbalanced systems, humus can be spent or replenished. The stable humus is vulnerable to many fungi, belonging to basidiomycetes and actinomycetes, especially when large amounts of cellulose or other carbohydrates are applied into the soil. Mycelium developing on carbohydrates with the help of extracellular phenoloxidase decompose the constant humus by 30% and more (personal information of Prof. G. Maier, Institute of Soil Science, Gettingen University, Germany).

Viewed as a whole, the formation and destruction of allelopathic toxins in soil is the part of general cycle of substances in soil and humus is the main constituent. Soil humus is mixture of numerous organic compounds, however, two types of polymer compounds (humic acids and polysaccharides) constitute about 90% and more of humus. Humus fraction soluble in alkali and precipitated by acids (at pH 2) is called humic acids. These substances are brown or nearly black. Substances which do not precipitate and remain in extract are called fulvic acids. They mainly consist of molecules of humic type acids with a relatively small molecular weight and of their polysaccharides. Humic acids and their close components are complex polymers of phenol units bound to amino acid chains, peptides and other organic components and can constitute 50-80% of the total soil humus (Gascho and Stevenson, 1968).

Soil polysaccharides are complex polymers from residues of sugars, which constitute 10-30% total humus (Metha *et al.*, 1961; Parsons and Tinsley, 1961; Swince *et al.* 1969; Martin and Haider, 1971). Soil polysaccharides can be derived directly from plants, for example, from mucilages formed by roots.

Mucous substances produced by maize roots contain polysaccharides, which after hydrolysis give 4 carbohydrates, including glucose and galacturonic acid (Samcevich and Romanov, 1975). The amount of mucilages produced by maize roots reaches 1250 m^3/ha, while that produced by wheat roots – 800 m^3/ha (Samcevich, 1966), which in terms of dry matter constitutes 0.4–0.5 t/ha of carbohydrates. However, a larger part of these mucilages is decomposed probably by bacteria. Generally the main producers of soil polysaccharides are obviously microorganisms and some extracellular processes. The majority of plant or microbial polysaccharides contains from one to three and very seldom four or five different monomer units. Polysaccharides isolated from soil as the main components, contain about 10 and more monomers and other sugars in small amounts. Apparently various polysaccharide fragments arising in the degradation processes are bound together into large molecules by extracellular enzymes present in soil. The remaining, approximately 10% of the total humus belongs to relatively mobile organic compounds of various origins. Other, often very active substances are nitrophenols.

W.A. Kalashnikov reported that lot of nitrophenols are present in soil humus layer and these compounds decreases the growth of crops in eroded fields, on the road sides and in humus deficient spots after removal of field surface by bulldozers or scrapers. Nitrophenols enter into the petrographic, components of soils, which is complex of saturated substances and maintained by alumisilicated soil particles (Bear and Thomas, 1964; 1965). Petrographic components are separated from soil by a distillation method with water steam and distinguished by large physiological activity, but has not been studied.

As we have seen, the nature of active substances, which can depress seed germination and plant growth, can decrease yield quality and change the composition of useful microflora for the worse, is very diverse. The phenolic substances with a higher quantity are most important and are beginning and end of cyclic transformation of humus compounds in soil. It is very interesting that the importance of so-called soil wax (lipophilic and volatile substances) has not been elucidated yet. Special places are occupied by the substances produced during the anaerobic fermentation in soil – alcohols, aldehydes, hydroxy acids, sulphides, hydrogen sulphide, ammonia. The specific metabolits of fungi, bacteria, actinomyceties, microscopic water plants and other co-organisms of higher plants contributes to soil toxicity.

Chapter 6

TOXINS : PHYSIOLOGICAL AND BIOCHEMICAL

P. Bogdan

Department of Allelopathy,
Central Botanical Garden, 1, Timirvzskaya Street, Kiev, Ukraine

> *"Cornfield is dried up by flax, is dried up by oat and is burnt by poppy filled with lethal dose. But sowing them at a year interval, it can be justified..".* Virgil, The Georgics, 35 year B.C. (Vergilus, 1971)

Activity of toxic substances to cause allelopathic soil sickness can begin only after their entering into the plants. Uptake of organic substances by plants has been found long ago. In the 18th century, A. Teier and K. Schprengel, the authors of "Humus Theory" of plant nutrition and later J. Liebig conducted experiments with plants fed by organic compounds. The ability of higher plant roots to absorb carbohydrates, amino acids, organic acids, phenolic compounds, antibiotics, growth substances, vitamins, synthetic insect-fungicidal preparations and other low molecular substances from the soil has been proved using the method of sterile cultures and radioactive isotopes (Grodzinsky, 1965). Mechanisms of uptake of organic substance are based on the ability to exchange metabolites with the neighbouring cells and environment. Such metabolic exchange takes place by transfer of molecules into protoplasmic membranes through plasmo- or ectodesma. Larger molecules can be taken up by pinocytosis (Salaiev, 1969).

In soil sickness, the most important is the entry of phenolic substances, which are intermediate products of synthesis and decomposition of humic acids. It has been proved that plant roots take up simple phenols and their glycosides. Large polymer molecules of humus do not penetrate cells. The break in phenolic ring in plant tissues occurs seldom - in the presence of methyl group in *para*-position. For this reason, phenols penetrating into the plants are most frequently glycosidized and thereby they are transferred into inactive compounds and do not participate in metabolism. In general, phenolic substances play an important role in (a) growth processes which has not been sufficiently investigated, (b) structure of cell walls (they enter into lignin composition), (c) the immunity and (d) many other manifestations of plant metabolism. Exogenous phenolic compounds occur most frequently as inhibitors of growth and other metabolic processes. Their activity is not specific and is determined by concentration (Kefeli, 1971). Terpenes or volatile oils (i.e. oil-soluble lipophilic or hydrophobic substances) are the second important group of substances taking part in allelopathy. There are many examples of their participation in soil toxicoses. A question arises about the mechanism of lipophilic substances penetration into plants. They are nearly insoluble in water, hence, their migration inside the multicellular organism by the usual pathways and participation in biochemical reactions taking place at the water phase are little real.

Müller (1965) dealing with the mechanism of volatile excretion activity of American sage species, had experimentally proved rapid and easy dissolution of terpenes studied by him (cineol, α- and β-pinen, camphor and others) in paraffin shavings and assumed, that terpenes also easily dissolve in waxes and oily cuticule covering the plant cells and move by diffusion in these oil layers penetrating the whole plant organism and forming the entire system.

Grodzinsky (1978) presented the mechanism of lipophilic substances in the following way. All ultrastructures of plant cells consist of membranes, of which the protein molecules are directed outside by their hydraulic layers and inside - by lipophilic layers. Since membranes are double-layered, lipophilic molecules form pores connected each other into entire system and formed a common lipophilic channel. This channel connects all cell structures having membranes and comes out to the cell surface into ectodesms and cuticule. Lipophilic molecules from the outside environment can penetrate into lipophilic channel to move by diffusion from cell to cell. In this way, molecules of lipophilic compounds present in the lipophilic channels can influence physico-chemical properties, which will be reflected later influence in the condition of cells and the entire plant.

Let us consider the effect of allelopathic soil toxicity on different aspects of plant metabolism. Soil sickness in horticulture and forestry, was displayed first by inhibiting the growth of shoots or seedlings, which after plantation continue to grow for some time on their own supplies and then they stop growing. In strong toxicosis, seedlings do not grow at all. Their leaves are twisted and get dry without falling down and 2-3 years after planting they die. Sometimes, soil sickness is displayed in the form of "dead zones" or spots. Leaf twisting in peach, the appearance of numerous offshoots, gum flow of fruit trees or the so-called apoplexy also frequently indicates soil sickness and the occurrence of toxic substances in soil.

Toxins influenced the grassy plants appearance through different morphological changes : first stoping of growth, thickening of the lower internodes, development of aerial roots and additional hairs. Stems and leafstalks develop anthocyanin colour. The leaves become chlorotic, necrotic and the accelerated disappearance of lower leaves.

One of the symptoms of shoot toxicosis is the loss of ability to orientate in space – stems are wilting, the first internodes are shortened, roots are horizontally spread and even come out on the soil surface and again deepen under the surface layer. It frequently happens that the first leaf makes a hole in coleoptile being still under the soil surface. These signs indicate the disturbance in balance between the growth substances and the plant wants "to go away" from the toxic zone, to avoid stronger toxin influences.

1. TOXIC PLANTS : COUCH GRASS

1.1. Agropyron repens

The influence of allelopathic factor adversely affects the initial stages of plant growth. Thus, water extracts from couch-grass rhizome and soil, on which couch-grass was grown, decreased seed germination and the length of alfalfa seedling (Ohman and Kommedahl, 1960). Aamisepp and Osvald (1962) reported that root excretions of couch-grass inhibited the seed germination and seedling development of rape, mustard, red clover, barley, mustard. Extract from couch-grass rhizome strongly inhibited the germination and growth of adult plants (Vicherkova, 1970). Bogdan (1971) found that water extracts from couch-grass rhizome (1:10) drastically inhibited the seed germination and seedling growth of vetch, pea, buckwheat, red clover, winter rape, flax and sunflower. The toxic effect of couch grass on cereal crops growth was weak in the beginning, but soon they also die.

The response of crops plants to couch-grass excretions under varying conditions is different. Rye, winter rape, vetch and oats show a weak response to couch-grass excretions, when their amount in crops is small and vice versa. Flax, red clover and strawberries suffer more from couch-grass, while wheat was moderately tolerant to couch-grass excretions.

The different soil microflora inhibited the seed germination of barley, pea, radish and cress and seedling growth in maize and cucumbers due to the accumulation of phytotoxic substance in the culture medium it (Berestecki and Patyka, 1976; Elisashvili, 1976).

Reduction in crop yield, when stubble and other harvesting residues are left in the field, can be determined by harmful substances produced by the soil microorganisms, which decompose these residues. In the first period of growth, in such fields with higher moisture content, there appear chlorotic plants with stopped growth. The reason is mass development of saprophytic fungi, *Penicillium urticae* Bainer (particularly under conditions of Nebraska State), which synthesizes an inhibitory patulin. The accumulation of physiologically active substances to toxic concentrations in the soil can deteriorate the process of mineral nutrition and reducing the nutrients absorbing ability of roots. Root excretions effects the entry of mineral elements into plants.

As a whole, the allelopathic influence increases the wastage of moisture, which increases the leaf succulence in some plants (Grümmer, 1957). This is probably because the toxic substances increased the protoplasm permeability, which ameliorates the transpiration, guttation and other losses of water.

It is evident that the index of leaf succulentness can be used as one of the signs of allelopathic soil sickness. Vicherkova (1970) showed that active substances from couch-grass rhizome reduced the transpiration, water content and osmotic pressure of cell sap, inhibited stoma opening in flax. She suggested that all these changes were secondary, whereas, primary is water deficit determining the limitation of water entrance through flax roots.

Toxic soil substances have unspecified effect on plant respiration and photosynthesis, i.e. at low temperature they can stimulate these processes, but at high temperatures they can inhibit and entirely suppress them. Our histological studies (Grodzinsky and Bogdan, 1975) have shown that allelopathically active substances penetrating the plants affect the activity of enzymatic systems, first intensifying their activity and later depressing it.

2. ENZYMES

2.1. Dehydrogenase

One of the first disturbances in Acceptor-plant metabolism, is the increase in unspecific dehydrogenase and succinate dehydrogenase in the phloem and parenchyma of vascular bundles. On the following days, the activity of these enzymes in the tissues of phloem and parenchyma decreased and were concentrated in the cavities of xylem vessels. This provides basis to consider that unspecific dehydrogenases and succinate-dehydrogenase enzymes are changing their activity as the first ones (Bogdan 1976).

Since the dehydrogenases and succinate dehydrogenases are peculiar to such important organelles as mitochondria, thus allelopathically active substances influence them (mitochondria). Our further studies have supported this hypothesis. We have noted that the activity of dehydrogenases increased for a short time. It is known that dehydrogenase activity is inhibited by quinones, which are the products of polyphenol oxidation. Polyphenols can be oxidized by polyphenol oxidase or peroxidase. Benzoquinones isolated from plants in certain concentrations can decrease the activity of peroxidase, catalase and polyphenoloxidase through interaction with their sulphhydryl groups. The activity of succinate dehydrogenase and ATP-ase are inhibited, the disturbance in the processes of oxidative phosphorylation and the synthesis of macroenergic bonds and as a result of it tissues obtain the indispensable energy by anaerobic respiration. A similar activity in succinate dehydrogenase is exerted by coumarin derivatives. In germinating wheat seeds, coumarin inhibits the peroxidase and changes its isoenzymatic complex.

2.2. Catalase

Catalase activity is inhibited by tannins, gallic and hydroxybenzoic acids. Various plant substances are inhibitors of malate dehydrogenase, glycerate dehydrogenase, tryptophan synthetase, lipase, hexokinase, amylase, invertase and others. In nature, the activity of oxidative and reducing enzymes of acceptor plants is influenced by kolines of unknown chemical origin or koline complexes. At low concentrations, kolines stimulate the enzyme activity to some extent, after increase in the concentration the activity of the same enzymes decreases or is completely inhibited (Shtsherbanovsky *et al.*, 1976). Juglone and plumbagin isolated from the plants increase polyphenoloxidase activity at certain concentrations. Triterpenic saponins activate the polyphenol oxidase stronger than steroidic saponins (Blagonravova and Shtsherba-nowsky, 1975).

The influence of simple phenols has been found on the process of enzymatic oxidation of ß-indole-3-acetic acid (IAA). Heterocyclic phenols with *orto-* and *para*-substituted hydroxyl groups are strong inhibitors of IAA oxidase. Ugrehelidze (1974) supposes that the inhibitory activity of phenols is determined by their oxidation into quinones and by interaction of quinones with IAA oxidase. A similar hypothesis was advanced by Stom (1975).

We found that healthy leaf tissues and young roots of many plants do not contain chlorogenic acid and other phenolic compounds with *orto*-substituted hydroxyl group. These compounds appear in sick plant tissues under the chemical activity of allelopathically active substances on them. Accumulation of large amount of phenolic compounds in plant tissues inactivates many biologically important enzymes and leads to their death (Bogdan, 1971, 1972). Thus, allelopathically active substances influence the plant organism by changing its enzymatic activity (induction or inhibition), which depends on the time of action and concentration of toxin.

Our histochemical studies have shown that the presence of kolines at toxic concentration in the extracts from clover, couch-grass and strawberries had unfavourable influence on the distribution of nucleic acids in acceptor-plants. In test plants, the amount of nucleic acids increased in comparison to control. Besides, the test plants have chromatin separation, which was particularly noticeable when treated with chloric acid and the tissues in phloem and in wood parenchyma nuclei were generally of large size.

With the further absorption of kolines by plant and with increase in pathological changes, the nuclei, chloroplasts and leucoplasts adsorbed dyes badly, which indicate decrease in the content of their nucleic acids. By the end of the experiment, nuclei, chloroplasts and leucoplasts were completely decomposed. A question arises, what has happened with nucleic acids in sick plants? Our studies suggest, that nucleic acids moved to places, which are most sensitive to toxic substances contained in the extracts or in sick soil. This conclusion was based on the observation, that there is a high content of nucleic acids in the free space and in intercellular spaces of many cells of bark and wood parenchyma, as well as in the interspaces of xylem walls (Fig. 11). As the plant sickness intensified, the amount of nucleic acids increased in the intercellular spaces and accumulated in large amount in the cavities of xylem vessels. Though kolines are able to cause changes in the content and localization of nucleic acids, this process does not refer to genetic apparatus of meristemic tissues on the side.

Interrelations of plant organisms in communities are based on interregulation, whereas chemical interaction can play a leading role in many cases. Chemical defence in microorganisms is accomplished by the production of specific substances influencing the regulation systems of the "enemy." It is quite possible that this principle is also true in chemical interactions of plants. The relations between the separate plant organisms are closer, when more hormonal regulation systems are included into their sphere of activity. As it is known, metabolism in a live cell is controlled by genetic mechanism of iso- and allosteric regulation of enzymatic activity and by a change in membrane structures and cytoplasmatic polyelectrolites.

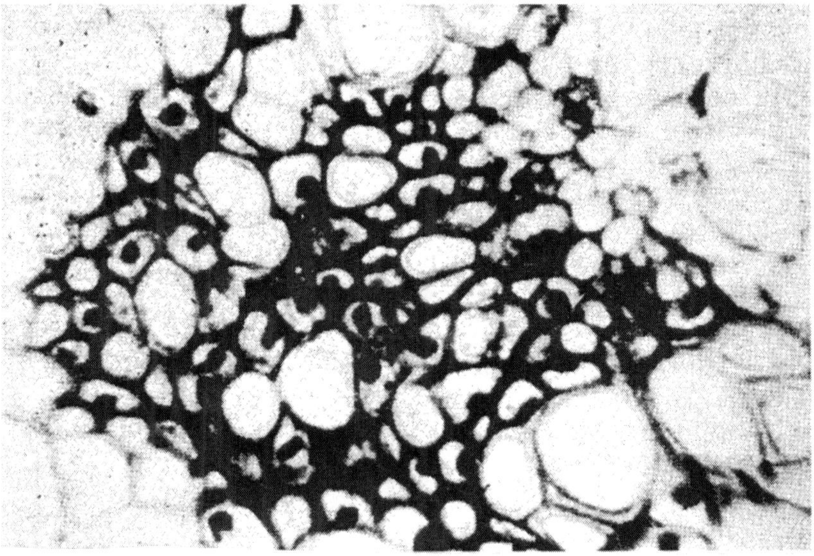

Fig. 11. Effects of extract from wild strawberry on localization of nucleic acids in the intercellular spaces of main oat bud treated for 4 days (40x10).

The composition of hormones connected with regulation of plant growth, includes auxins, gibberellins similar substances, kinins, growth inhibitors and also enzymatic systems participating in phytohormone synthesis and destruction. The work of genetic cell apparatus, membrane properties or permeability of cellular structures is under the influence of hormones. The activity of IAA on the growth is inhibited by different substances, for instance quinones and phenols, which can form a complex with IAA or activate its destruction. On the other side, the action of IAA is intensified by some phenols and vitamins, which are not considered in plant excretions. Besides the activity on phytohormone system, the excretion products of some plants can influence the work of genetic apparatus of other plants, on enzyme activity, on the properties of membrane structures, semipermeability of cells, etc. or on epiphitic microflora, which is very important for the balance of phytohormones in plant tissues. Disturbance of intracellular regulation, structure of membranes and cytoplasm polyelectrolytes will lead to a drastic decrease in metabolism of plant organism.

Bukolova (1969) studied the effect of various concentrations of water extracts from above-ground plant parts and roots of pig weed, canada thisle and sonchus on mitotic activity of meristematic cells of rye, wheat and pea roots. Most of the tested extracts depressed the germination of crops plants, which delayed the rootlet appearance and decreased their length. These symptoms are connected with the disturbance of mitotic apparatus. Kolines reduced the mitotic activity and cause many abnormalities in chromosomes. High concentrations of weeds extracts (1:10, 1:20) strongly inhibited the activity in mitotic apparatus of rootlets until the suppression of mitotic cycle. Lower concentration of kolines disturbed the mitotic activity and caused chromosome agglutination, their thickening and shortening and incorrect divergence at anaphase bridges. Concentration and time of koline action on plants are determining factors in this process; species-specificity of plant does not play a special role here because koline effect is unspecific.

The influence of kolines is reflected on the process of tissue differentiation. The most marked disturbance in differentiation process of tissues was observed in wheat growing in a sand culture with weed residues (2.5% of sand mass). The plants were under the influence of the above-ground part of sonchus decomposing in the sand for 30 days. Under these conditions, wheat roots turned brown and though they looked healthy but the signs of the initial stage of tissue decomposition were found under the microscope. The water soluble and volatile excretions reduced the diameter of the central cylinder, when the bark parenchyma cells were larger in size. Metaxylem vessels had brown walls. Miazmines from the roots of pigweed had an opposite effect. The central cylinder of wheat roots increased due to the increase in diameter of xylem vessels. Some vessels were deformed and cavities appeared. Thus miazmines of sonchus had an analogical effect on the root system of rye. On the contrary, the development of the cress root system was stimulated by the excretions of decomposing residues of weed plants, which reflected in the development of the conductance root system and increased the amount of secondary xylem in comparison to the conductance system of control plants. The spread of bark parenchyma cells was very weak and the epidermis was easily exfoliated.

The koline application to 2-3 days-old seedlings of wheat and pea with the rootlets of 20-25 mm length, caused peculiar changes in their roots. It influenced the mitosis and germinating seeds, the growth continued but plants were drastically depressed. The epidermis and bark parenchyma were damaged, the root tip, the meristem zone and absorbing zone suffered the most. The protoplasm in these cells plasmolysed and the all walls turned brown. In root hairs zone, only epidermis and some layers of parenchyma cells adjacent to epidermis were damaged. The degree of damage depended on the concentration of extracts (Bukolova, 1969, 1971).

Changes in mitotic apparatus of plants occur not only under the influence of kolines of plant origin, but also the metabolites of soil microflora have similar effects (Elisashvili, 1976). Besides the disturbances in mitotic apparatus, microflora toxins also cause abnormalities in the nucleus : both sticking together of chromosome in anaphase, lysis and polymorphism.

Our studies have shown that allelopathically active substances present in the environment of germinating seeds or seedlings affect the structures responsible for hereditary cell properties. Chasovenna (1972) reported that the physiologically weakened plants become more sensitive to volatile phytoncides than solid ones. Weakened plants were found to lose turgor earlier, to be irreversibly senescing, to go quickly in lethal state, than normally developed plants. In comparison to control plants, those treated with volatile phytoncides in the beginning showed more intensive cell division and stoma formation in epidermal part of leaves and in cotyledons.

The semi-saprophytic microorganisms develop on the roots damaged by insects or nematodes and they often produce active substances, which accelerate the plant death (Peryov and Mirzayev, 1971).

3. STUDIES OF TOXIC PLANTS

3.1. Couch grass

Under sterile conditions, the permitted amounts of toxic substances, display no action but weaken the test plant and decrease its immunity to fungal diseases. Lastuvka (1961) found that agropyren application under sterile conditions had no influence on wheat, but the same dose of toxic substances of couch-grass in soil culture caused infection of wheat with *Stemphylium ilicis*, which usually never infects wheat. Generally this secondary infection with diseases is considered as the cause of soil sickness, reduces the crop productivity and causes the death of plant, but the initial cause remains unknown.

The yellow-brown substance accumulated in the roots of wheat grown with couch-grass led to the identification of agropyren (Lastuvka, 1961). He gave the following characteristics of agropyron: part of the brown colour in thin root cuts was dissolved by ether, ethanol, acetone and chloroform. It intensively absorbed dye-stuffs (safranin, eozine, aniline dye and others). Refraction index in yellow-brown matter is higher than that in water and glycerin. This substance was extracted by ether from the powdered wheat roots. When heated in the presence of oxygen, it changed into hard brown substance. It react successfully with nitric acid and when heated with alkali, it turned brown looking like resin. All the properties are similar to those described by Treibs (1947, 1959). Lastuvka underlines that this substance is in active condition and after penetration into the wheat roots its properties do not change. It moves through the vessel-conducting bundles and reaches the couch-grass roots and leaves.

For deeper knowledge about the mechanism of couch-grass influence on the anatomical structure of cultivated plants, we have also conducted studies on the crops growing under the influence of couch-grass in natural field conditions and lab conditions. In laboratory studies, roots of crop plants were dipped in water extracts (1 : 10, 1 : 20 and 1 : 40 ratio) of dry couch-grass rhizomes. Crops used in the experiments were: strawberries, cv. 'Kievskaya runniaya', red clover, buckwheat cv. 'Krylataya 70', vetch cv. 'Belocerkovskaya 27', winter rape cv. 'Mytnicky 1', sunflower cv. 'Masliczny', oat cv. 'Lgosky 1023' and winter wheat cvs. 'Mironovskaya 808' and 'Bezostaya 1'.

Microscopically investigated stems of rape, vetch, oat and sunflower showed the presence of brown mass (homogeneous or lumpy consistency), in the cavities of xylem vessels above the root collar and frequently even in leafstalks or in leaf blades near the first leaf. In stems of young plants the browning of xylem walls occur frequently, when the plants are treated for longtime with the extracts and the browning also reaches the leafstalks. The older plants were more resistant to couch-grass excretions and the browning and corking of xylem vessel.

I. Strawberry

The strawberry is most sensitive to couch-grass excretions, its leaf senescence is observed soon after placing the plant in water extractions from couch-grass. The bark parenchyma and peripheral part of the central cylinder die as seen in the cuts made in young roots. There are no changes in the lower part of the stem and the older the plant, the lesser is damage. In the middle part of the stem, only the walls of parenchyma cells surrounding the conducting pathways turn brown. Pathological changes in the upper part of the stem are of different nature. Here, the parenchyma surrounding vessels are healthy and only the walls of xylem vessels leading to stalks and old leaves turn brown. Xylem leading to young or developing leaves turns brown in the end. Brown matter corking the xylem vessel cavities is found in places of fruit fastening to calyx, whereas, browning of only xylem walls occurs in inflorescence.

II. Red clover

In red clover, the localization of pathological changes caused by couch-grass influence also depends on the plant age. In stems, the brown matter in xylem is concentrated in the nodes above the root collar.

III. Buckwheat

In laboratory and under field conditions, the corking of buckwheat xylem cavities and their wall browning occur near the nodes and leafstalks. Under field conditions, buckwheat was heavily depressed than control and at the beginning of flowering, the anthocyanins appeared in leaves and soon the leaves got dry and no yield was obtained from that plot. Many tumours were found on the roots. Under field conditions, in the plot, where a small population of couch-grass was introduced in rows at crop sowing there was no, xylem browning and the crop suppressed the couch-grass.

IV. Oats

In oats, growing on a plot with a very large amount of couch-grass (on the dug over turf), xylem vessel corking with a brown matter was observed in the node of tillers and in the first node of stem. In many plants, additional rootlets were formed near the first node. Oats growth was depressed, lower leaves were chlorotic and some plants were infected with white spike disease.

V. Wheat

In the plots, where couch-grass was added in rows at wheat sowing, in the beginning the growth was not influenced. At the milk stage, the leaves of the susceptible cv. 'Mironovskaja 808' and resistant cv. 'Bezostaya 1' were strongly infected with brown leaf rust, whereas, wheat leaves without couch-grass treatment remained healthy despite a very strong infection. The cause of the infection of resistant wheat cultivar is the decrease in its defence mechanism due to couch-grass excretions. While, at the dough stage, the brown matter accumulated in the walls of xylem vessels in the roots, in the node of tillers and in the first node of stem, i.e. the oldest part. Above the node (in the young intercalary zone), the vessel browning and corkining do not occur. In nodes, part of the bundles turned brown and part remained healthy. It was found that the bundles leading to the nearest leaf, turned brown, whereas, those leading to the upper part of stem remain undamaged. The most brown mass in the xylem was in the area of first leaf.

3.2. Clover

We decided to elucidate, whether there is corking of the vascular bundles in plants in the typical cases of soil sickness. One is red clover, whose stand become sparse and plants dies by the end of the second and beginning of the third year. Many studies had been done to know the reasons of a premature clover death since the start of 19th century. Some investigators connected this event with soil sickness, stipulating it by nutrient phosphorus and potassium deficiency. Kossovitsh (1905) showed that sufficient fertilization also did not remove soil sickness completely. Clover failure is observed also on the rich soils, not suffering from the nutrient deficiency and are completely suitable for clover growing.

Clover roots are damaged by multiple insects, as a result of which pathogenic microflora has a free access to plant tissue, which is considered to be the main reason of clover death (Nikitina and Nagovicina, 1972; Nagovicina and Nikitina, 1975). In the first and second year old plant roots, the presence of *Fusarium* was in 32% and that of bacteria in 54% plants, which concluded that disease agents results in clover failure.

The problem of clover failure under conditions of Ukraine was studied by Beltiukova and co-workers (1959). They consider that the reason of clover failure was bacteria, which entered the plant tissues through wounds, splits and cut down stems. The symptoms of disease were very different. Brown spots appear on the bark of damaged plants, gradually they cover the entire area of root neck. Vessel bundles and the core turn brown. Microscopic studies revealed the presence of a large number of bacteria in both, vessels and tissues near them. The damaged roots decay, turn yellow and die. After decay of the main root, adventive roots are formed in the area of the root collar, but the plant cannot survive with their help for long time. Bacteria isolated by Beltiukova and co-workers are the cause of root decay only at the first stage of the pathological process.

In the majority of the studies conducted, the initial reason of clover failure is plant weakeness caused by the unfavourable conditions or by their reduced vigour. Krasilnikov and Garkina (1946) found that toxic substances suppressing the nodule bacteria, nitrobacter and clover plants themselves accumulate in the clover sick soil. The application of small pieces of clover roots and stems to soil, strongly inhibited the growth of clover plants. The Japanese investigators studying the reasons of clover soil sickness, reported that leaves and stems of red clover (*Trifolium pratense*) contain nine inhibitors of growth (ononin, diazein, diazein-7-glucoside, formononentin, genistein, biogenic A, biogenic A glucoside and 5-malonate trifoliridzin). They extracted the sick soil by 50% water methanol and after dividing it into neutral, acid and ethyl acetate fractions and after chromatography, an oily substance was obtained. It contained the following phenolic acids : *p*-methoxy-benzoic, salicylic and p-hydroxybenzoic at 1, 20 and 5 mg/kg of soil, respectively. Isoflavonoids are not found in the soil, but they are undoubtedly a source of the additional phenolic compounds, which are produced after their decomposition. Nutrient solution of clover water culture was found to contain the following acids in a free state: bioxanin A and formononetin, *p*-methoxybenzoic, *p*-phenylacetic and cinnamic acids. A simultaneously studied sample of the same soil under Italian ryegrass, was found to contain traces of *p*-hydroxy-benzoic and 2,4-dihydroxybenzoic acids. Hence, very large quantity of flavonoids and their high concentration can partially explain the reason of soil sickness of clover. They have attempted to show similar inhibitors in the clover sick soil. However, only the acid fraction separated from the methanol extract was toxic and the expected compounds were not found in it. Thus investigators concluded that soil sickness of clover mainly depends on the accumulation of such phenolic compounds, as isoflavonoids and their derivatives, which are excreted by red clover during the vegetative growth and after their release they are converted into other compounds by living plants.

Despite the enormous research, clover failure from allelopathic point of view is poorly presented in the literature. We have tried to fill this gap with our investigations. We studied the reasons of clover failure in the field trials and laboratory tests for 3 years. Xylem turning brown in the first year of growth was seldom occurred, in the second year (after the first hay harvest) about half of the investigated plants had browned roots, and after the first hay harvest on the third year clover culture died completely. Microscopic investigations of roots of the third year of plant life showed that almost the entire xylem was filled with the brown matter. It is quite probable that clover dies because it appears to be unable to transport water with dissolved substances to the above-ground parts of plant due to mechanical obstacles i.e. brown matter in vessels. The reasons of the brown matter occurrence can be different. Under laboratory conditions, clover seeds placed in Petri dishes on filter paper moistened by the extract of above-ground matter of clover in 1 : 5 concentration did not germinate, while in 1 : 10 concentration the germination of seeds was poor and their seedling growth was poor in comparison to the control (water). Under laboratory conditions we performed also model experiments. Roots of healthy plants was taken from the first vegetative growth and placed into the extracts prepared from fresh or dry shoots of clover. Inhibition, senescence of leaves and their death were observed on the second-third day of the experiment and all plants died or 6th day. Microscopic studies show that under the influence of physiologically active substances contained in the extracts, pathological changes occured in xylem identical to those taking place in plants from the field plots, where clover was grown for 2-3 years.

Our investigations show that the initial cause of clover failure is the accumulation of physiologically active substances harmful to plants in soii and conducive to the development of specific microflora. Physiologically active substances are gradually absorbed by plant and they poison it. The weakened organism is more susceptible to any infection or is more damaged by unfavourable winter conditions. In our opinion, occurrence of fungal or bacterial microflora can be a secondary event. On the basis of our investigations, we suggest that the reason of clover weakening can be allelopathically active substances accumulated in soil, which the gradually disturb the metabolism, leading to wall browning and corking the cavities of xylem vessels with the brown matter.

3.3. Strawberry

Soil sickness is also noted in strawberry monoculture, which weakness the plants, and crop yield decreases and runners are not formed. In the fourth – fifth year, most of the plants die. In old plants, large amount of leaves die. Leaves fallen on the soil gradually decompose due to precipitation (rainfall etc.) and releases saprolines (Grodzinsky, 1965), which effect the growth and development of strawberry. The amount of saprolines increases with each subsequent dieing of leaves and roots, which intensifies the soil toxicity and weakens the plant defence system reducing its crop yields. The leaf dying and its decomposi-tion is one of the major reasons of strawberry plants death.

3.3.1. *Anatomical changes* : To know the anatomical changes in different organs of strawberry, we took plants of the first to fifth years of growth under field conditions. Old dying leaves were used to prepare water extract (1:10 ratio). Plants of different ages were maintained in the water extract at different vegetative stages. The experiment lasted for 7-8 days. Microscopic sections were made from fresh roots, stems and leafstalks to examine them under a microscope.

After 3-4 day of plants growth in the prepared extract, a strong wilting of lower leaves was observed. No special changes occurred in the roots during the first days. Cells of bark parenchyma and peripheral part of the central cylinder in the area of root collar turned brown when the exposure to extract was enhanced. Young formed rootlets showed strong allelopathic influence. With increase in the age of roots after periderm formation, their sensitivity becomes weaker to allelopatically active substances and no pathological changes were found. No anatomical changes were noted in the lower part of strawberry stem. In the middle part of the stem, sometimes browning of walls and parenchyma cell cytoplasm was noticed. Parenchyma at the stem apex adjacent to the conducting bundle looked normal, but xylem walls and cells of wood parenchyma present in vascular bundle became brown and sometimes this browning enters into leafstalk and leaf. The changes in xylem resulted in rapid leaf wilting.

3.3.2. ***Pathological changes*** : In summer months, outside symptoms of wilting were the same. No pathological changes were found in xylem and in leaf stalks. Only in the upper stem, the walls become brown and the cavities of xylem vessels going to old leaves are corked with brown matter. The parenchyma surrounding the vascular bundles did not turn brown in these plants.

Pathological changes were not found in young strawberry plants of the first and second year of growing on non-sick soil under field conditions. In stems, in the places of old leafstalks, brown matter was localized in the cavities of xylem vessels. In dieing old plants, at the stem apex, the cavities of nearly all the vessels going to leafstalks of young and old leaves appeared to be filled with the brown matter. All vessels going to young and old leaves were also corked in young plants on sick soil, which subsequently led to the turning brown of the entire stem apex and later of the entire plant.

Thus the reason of leaf death in aging plants was corking of xylem vessel cavities in the area of leafstalk separation from the stem for a distance of 1–3 mm. Similar xylem damage took place also in the plants, when they were maintained in the extracts from old leaves under laboratory conditions. In our opinion this event explains the death of aging leaves and consequently of entire plant. Under the influence of accumulated allelopatically acive substances in soil, the metabolism was disturbed, hence, the cavities of xylem vessels in strawberry stem were filled with the brown matter. Because of disturbance in vessel conductivity, these plants became too weak and were severely infected by fungi and died soon.

I. ***Brown matter*** : As we have mentioned above, the similar pathological changes were noted by Lastuvka in spring wheat grown with couch-grass. In his opinion, the brown matter localizing in wheat tissues is agropiren excreted by couch-grass, which dissolves in organic solvents (Treibs, 1947, 1959).

To learn the dissolubility of brown matter we examined the thinnest microscopic sections prepared from the stems of rape, oat, barley, strawberry, vetch, phlox and wheat plants maintained for 8 days in different extracts. We boiled these sections for half an hour in water, left them for four-five days in ethyl acetate, acetic, sulphuric and hydrochloric acids, methanol, water, ether, chloroform, xylol, strong solution of alkali, ammonia, glycerin, 10% hydrogen peroxide, Carnoix mixture. Brown matter was not completely extracted from the vessel cavities. In a strong solution of alkali, it turned light and after boiling in the same solution for 30 minutes the microscopic sections became completely discoloured. Concentrated sulphuric and hydrochloric acids did not destroy the browned tissue. In entire plants and in stem fragments 5-7 cm long, the brown matter corking the vessels could disappear, if the plants maintained in the extracts were left in water for some days.

In whole plants of clover, maintained for 8 days in the extract from couch-grass, clover, nettle, the root ends were slightly cut, and each plant with noticeably browned vessels were put in water. After 7 days, young leaves and roots began to grow in these plants in water, especially intensively in clover which was preliminarily maintained in extracts from nettle. It appeared that brown matter nearly disappeared in all roots, both in cavities and vessel walls. Only single vessels remained to be corked. In some clover plants, in which new leaves and additional roots did not grow, brown matter remained completely in all vessel cavities, i.e. it did not disappear.

This phenomenon may be apparently explained in this way: a normal course of biochemical processes in the thinnest tissue sections is disturbed and substances causing vessel corking are not washed out by water. Whereas in tissue fragments or in entire plants after the end of toxin action, biochemical processes are not completely disturbed, due to which plants can produce enzymes promoting brown matter dissolution and vessel release from corking. If very deep changes occurred in the brown matter, or if the plant died, no dissolution of brown matter was observed.

The investigations done on the chemical nature of brown matter causing browning and corking of xylem vessel cavities, show that it is not agropyrene or any other extraction substance excreted by couch-grass or by other plants. We consider that this matter is produced in the acceptor-plant under the influence of couch-grass or other physiologically active substances.

Our histological studies have found that in certain periods of the brown matter formation in xylem vessels, physiologically damaged by active substances, pectin compounds are formed (Fig. 12)

(Grodzinsky and Bogdan, 1972). The plant enzymes can convert pectin into soluble pectins, as a result of which we observed xylem vessel liberation from brown matter, if they are transferred into pure water for some time.

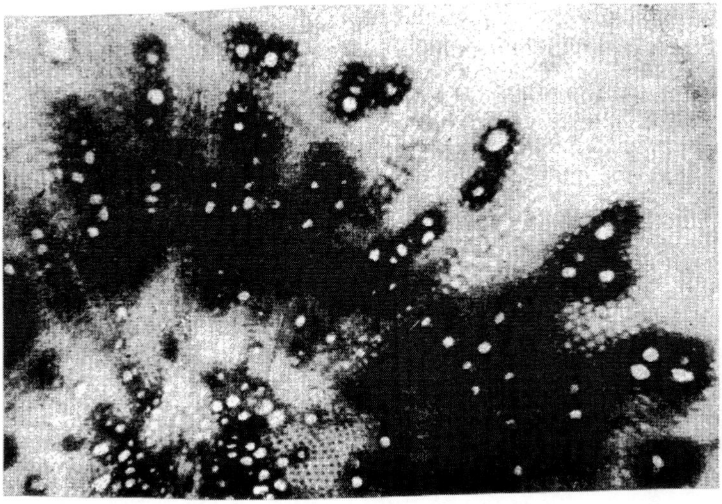

Fig. 12. Effects of extract from wild strawberry on localization of pectin compounds in cell walls (of turning brown vessels) & surrounding parenchyma cells of rape root wood treated for 5 days (8 x 10, 40 x 10).

In different periods of corking formation in xylem vessels, the brown matter has different pH index (Bogdan, 1974). In the initial stages of brown matter formation, the pH decreases and the medium become acidic. However, in the subsequent periods of corking formation, pH becomes neutral and alkaline. The pH plays an important role in the enzymatic processes, because enzymes are more active at certain pH. When pH declines from the optimal value and become acidic or alkaline, the activity of enzymes becomes weak or they are completely inactivated. Thus, the brown matter during the initial periods of its formation has slightly acidic medium, which promotes the activation of enzymes such as unspecific dehydrogenase, succinate dehydrogenase, peroxidase, ATP-ase (Fig. 13). Because pH value of the medium changes soon to neutral and alkaline in gum-similar substance, this is reflected in the activity of the studied enzymes. Our histological investigations have shown that corking in the same vessel have unequal enzyme activity. In some xylem vessels or even in part of the vessel, the enzyme can be active, whereas in the other part of the vessel its activity is suppressed. This indicates heterogeneity of chemical composition of the substance corking the vessels. Localization of enzymes in the cavities of xylem vessels are indicative of high oxidizing potential (Grodzinsky and Bogdan, 1975). Activation of dehydrogenases, succinate dehydrogenase, and peroxidases promotes the formation of corking, whereas, cytochrome oxidase does not participate in this process (Bogdan, 1974; 1977).

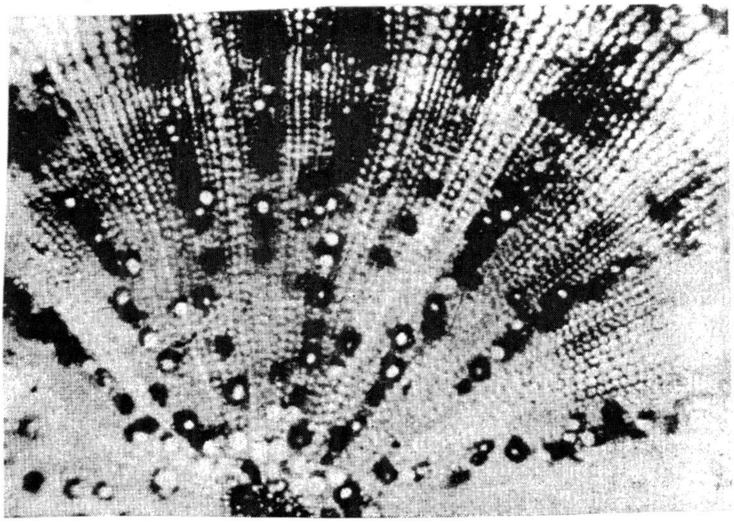

Fig. 13. Effects of extract from clover on ATPase activity in the vessels of rape root xylem treated for 5 days (8 x 10).

4. KOLINES

4.1. Ascorbic acid

Medium acidifying the sick parts of tissue creates favourable conditions conducive to the accumulation of the reduced form of ascorbic acid. Fulfilling the function of hydrogen carrier, ascorbic acid is closely related to the entire system of enzymes involved in the respiration process. Ascorbic acid is usually present in places of increased metabolic activity. Healthy tissues of the plants studied by us, contained ascorbic acid in comparatively small amounts, whereas in plants under the influence of toxic concentrations of physiologically active substances, the pathological process starting in damaged vessels is accompanied by accumulation of the reduced form of ascorbic acid. Therefore, ascorbic acid participates in the formation of substance similar to brown gum (Fig. 14) (Grodzinsky and Bogdan, 1973).

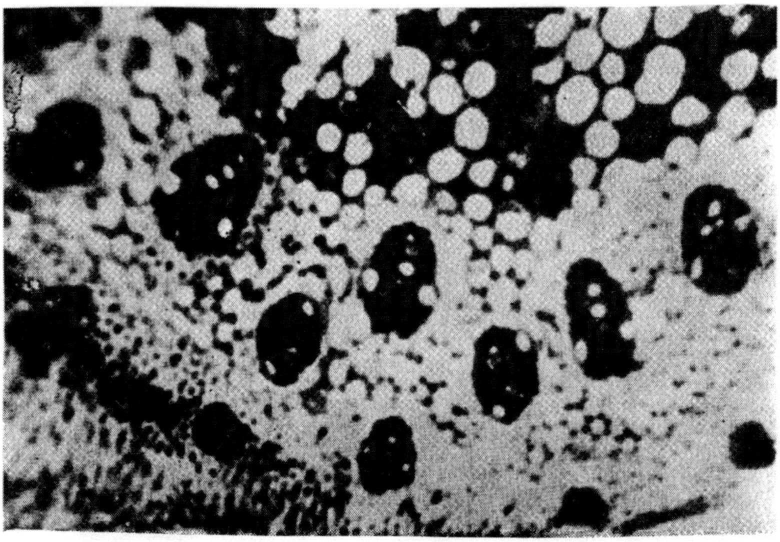

Fig. 14. Effects of extract from quack grass on localization of reduced form of ascorbic acid in main buds of oat stem treated for 6 days (20 x 10).

Nevertheless among all the above substances mentioned by us, the reduced form of ascorbic acid in gum-similar substance is probably gradually oxidizing or is bound with proteins, transforming into a homogenous mass, which does not react positively with silver nitrite, for the presence of ascorbic acid.

According to literature, ascorbic acid is oxidised only when the entire glutathione is almost oxidized. Presence of the reduced form of glutathione, prevents enzymatic oxidation of the ascorbic acid, i.e. in the presence of glutathione, dehydroascorbic acid is energetically reduced. Glutathione like ascorbic acid can exist in two easily reversible forms – reduced and oxidized. A comparatively easy oxidation of reduced form of glutathione by oxygen into simple sulphhydryl compounds indicate that these substances play an important role in oxygen activation during the plant respiration. Thus functional activity of cells depends on the amount of oxidized and reduced forms of glutathione and the ascorbic acid in them.

Sulphhydryl groups (Fig. 15) side by side with the reduced form of ascorbic acid participate in complicated processes of gum-similar substance transformation in xylem vessels (Grodzinski and Bogdan, 1973). Besides, the disulphide compounds (Fig. 16) are also found in gum-similar substance. This indicates oxidation of glutathione and ascorbic acid. The oxidized forms of glutathione and ascorbic acid are inhibitors, while, reduced forms are activators of proteolytic enzymes.

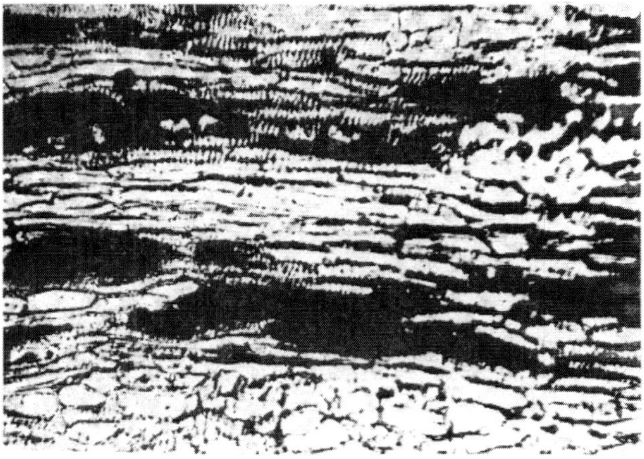

Fig. 15. Effects of extract from wild strawberry on localization of sulphydryl groups in brown mass corking xylem vessels of clover root treated for 10 days (20 x 10).

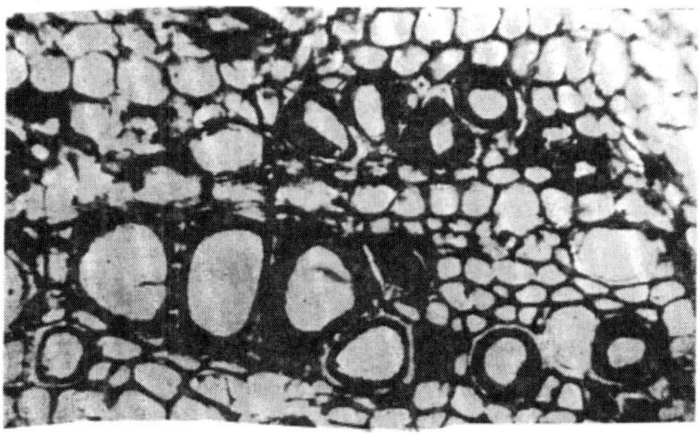

Fig. 16. Effects of extract from quack grass on localization of disulphite groups in intercellular spaces and in the vessels of clover xylem root treated for 6 days (40 x 10).

Owing to the presence of phenolic substances (chlorogenic acid) and high oxidation potential of brown mass, gradually the deamination of amino acids takes place to ammonia (Fig. 17). Ammonia creates favourable conditions for further conversions of gum-similar substance.

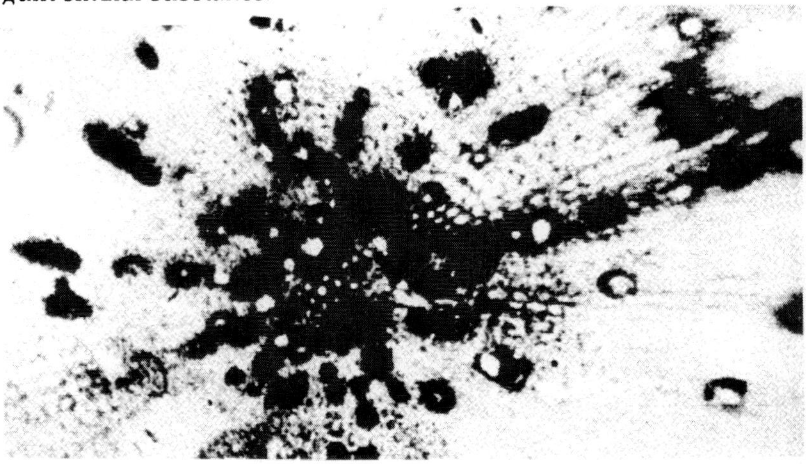

Fig. 17. Accumulation of ammonia in xylem vessels and wood parenchyma at the beginning of sickness of rape root kept in extract from clover.

4.2. Lignin

As per Manska (1957), at the atrophy of living cells in the process of lignification, oxidizing processes predominates over the reducing ones and the products of polyphenol oxidation begin to accumulate as lignins. In turn, the products of lignin accumulation disturbs the respiratory apparatus. The oxidation products convert into lignin after condensation. Precursors of lignin can be polysaccharides – cellulose, pectin substances and water soluble, carbohydrates. Besides that, precursors of lignin are coniferyl, feruoyl, sinapic, p-hydroxycinnamic alcohols convert into hydroxyconiferyl alcohol and the last, condensing into lignin. Lignification process can be controlled by auxins, restraining a premature influence of peroxidases on metabolites necessary for this process. The amount of coniferyl alcohol available for peroxidase action can be controlled by glutathione and ascorbic acid.

Lignin is formed at certain stage of pathological conversions of brown matter in the damaged vessels (Fig. 18). However, the process of corking in the vessels is dynamic and lignin after its formation is obviously converted soon into other compounds, thus, it is seldom observed in the stem and root sections of plants kept in the extracts for long time. Unequal colouration of brown matter indicates complexity in its chemical composition (Grodzinsky and Bogdan, 1972).

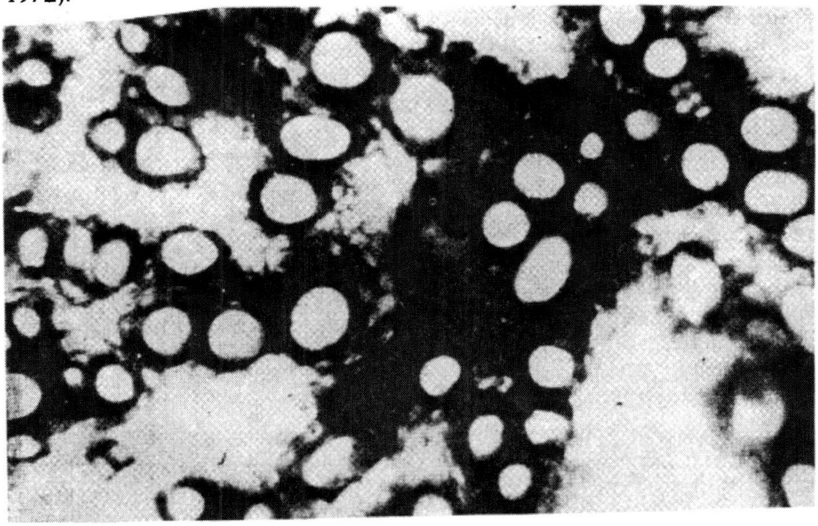

Fig. 18 Lignin accumulation in damaged vessels of 2-years-old clover (20x10).

Using histochemical methods, the main amino acids (Fig. 19), proteins and nucleic acids were found in brown matter. Localization of proteins and nucleic acids (Fig. 20) in the brown matter of damaged vessels indicates that the most important substances in genetic and functional aspect, i.e. amino compounds and amino acids, are transported to the place of toxin penetration. Besides the plant tränsports the damaged mitochondria vessels (Fig. 21), in whose membranes all the important enzymes for tissue respiration are concentrated.

Our investigations suggest that there is a site near the xylem vessel cavities and parenchyma cells, where toxic substances entering the plant are inactivated with direct participation of mitochondria. Due to oxidizing processes, the energy reserves are exhausted and mitochondrias die or are destroyed, then they themselves forms the corking substance for the vessels. The brown corking matter is complex of substances resulting from the activity of mitochondria mobilized from the whole plant for its defence, as well as decomposition or oxidation products of dead mitochondria.

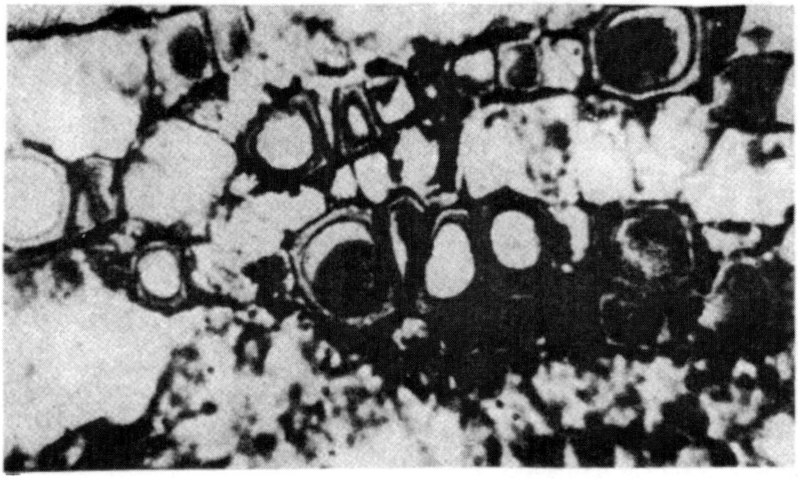

Fig. 19. Effects of extract from clover on localization of basic proteins in intercellular spaces & in the rape xylem vessels treated for 9 days (40x10)

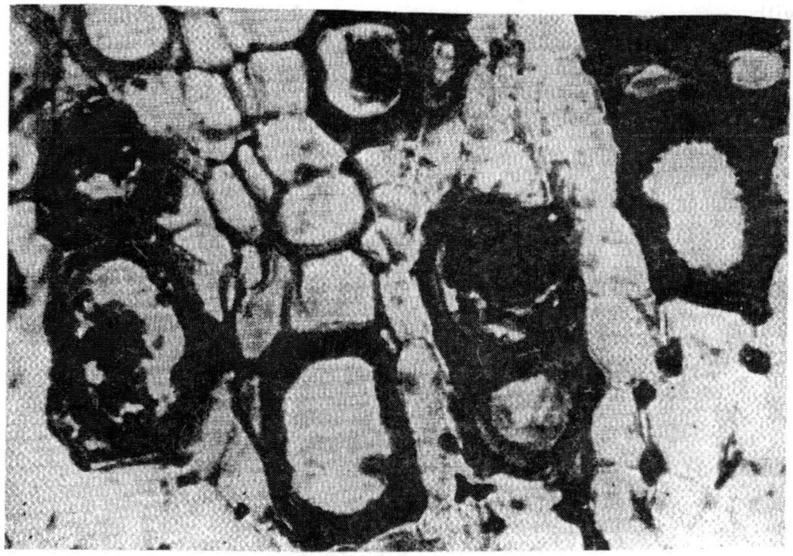

Fig. 20. Effects of extract from quack grass on localization of nucleic acids in intercellular spaces and in the xylem vessels of rape stem treated for 9 days (40 x 20).

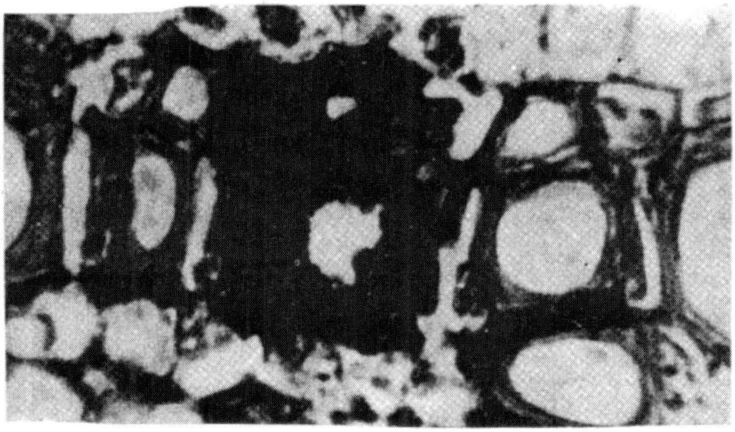

Fig. 21. Effects of extract from clover on localization of mitochondria inside damaged xylem vessels of rape stem treated for 9 days (40 x 20).

The final product of metabolic conversions of gum-similar substance in the damaged vessels is the formation of melanines (Fig. 22). Melanines are dark coloured condensation artefacts of nitrogen containing products, which occurs in the interaction of amino acids with sugars, uronic acids, ascorbic acid and phenolic substances in alkaline medium. Melanines are strongly polymerized substances, extremely stable to the action of solvents as well as to the action of chemical and biological factors.

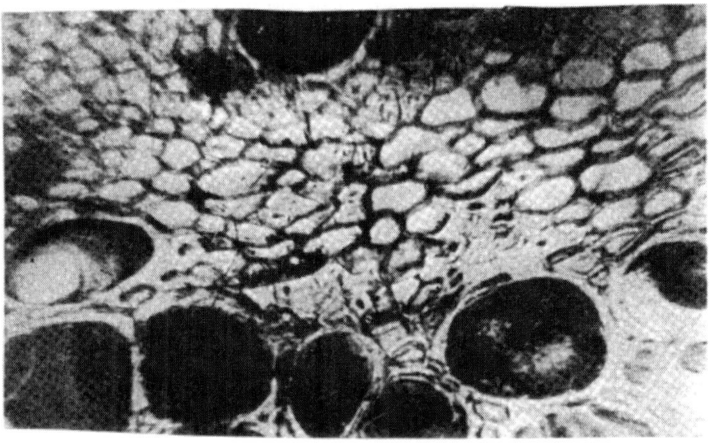

Fig. 22. Effects of nutrient solution of Helrich for plant regeneration on localization of melanins in brown mass corking xylem vessels of clover roots treated for 2 months in (40 x 10).

Therefore gum-similar substance causing xylem wall turning brown and corking of its cavities represents the formation of an extremely complicated dynamic compound. Brown gum-similar substance arising in the cavities of xylem vessels under the influence of toxic concentrations of allelopathic active substances is the product of defense response of plants.

Hence, the nature of the influence of allelopathically active substances on plant organism is very complicated, diverse and depends on many factors (plant age, its specific properties, intensity and time of action, environmental conditions and concentration of substances entering the plant). All these factors can determine the both, positive and negative affect.

Allelopathically active substances in toxic concentrations exert a negative influence on (a) germination energy, (b) germinating ability, (c) growth and (d) productivity. Under the influence of kolines, the absorbing ability of roots is depressed, which inhibits the entry of individual nutrients into the plant. This can also delay the growth and causes the death of plants organism.

Root excretions changes the water relationships in soil and in plant, to disturb transpiration, respiration, photosynthetic activity, chloroplast activity and carbon assimilation. Kolines affects the (a) structural organization of protoplast, (b) constitution of its organelles, (c) submolecular chemistry in plants (d) disorganise the enzymatic activity, (e) influence the substances proteins, amino acids, nucleic acids, phenolic compounds, carbohydrates important in biochemical interaction and (f) Chromosomes of epistemic tissue, which may change the gene apparatus of cells.

5. CONCLUSION

Toxic concentrations of kolines (a) exert influence on anatomical structure of plants, (b) leads to pathological changes in xylem part of the conducting system, corking it by brown gum-like substance and (c) forms brown accumulations in vessels of xylem as unspecific defense response of plant to different unfavourable factors of the environment.

Brown substance in the xylem vessels is the product of defense response of plant, for its production energy pool of plant is wasted. Easily influenced substances of leaves and stems are dissolved and transformed into compounds, which are directed to the places of the heaviest tissue damages – xylem vessels. Toxic substances are inactivated by their binding or oxidation. Mitochondria enters into the xylem vessels to form the high oxidizing potential, under whose influence the mitochondria serve as a basis for further metabolism into brown substance and its final products are melanine-like substances.

As the plant responds to entry of toxic concentrations of allelopathically active substances, by defense response using reserved easily influenced substances, Hence, the defense functions of plant should be increased for the successful control of soil sickness by applying nutrients to increase the vigour of plants and its redox potential. The variability in crop varieties or crops is also important here.

In future, the allelopathic potential of the plant environment surroundings should be reduced by (a) inactivation of toxic substances through chemical methods, (b) selection of tolerant plants, capable to metabolize, bound or render them harmless by their own excretion products. For example, biological methods can be used to control couch-grass. Vetch or oat is sown and after their harvest sow winter rape to clean the field from couch-grass within one year.

Chapter 7

MICROORGANISMS : ROLE IN SOIL SICKNESS AND SOIL FERTILITY

E.A. Golovko

Department of Allelopathy,
Central Botanical Garden, 1, Timirvzskaya Street, Kiev, Ukraine

"Main peculiarity of soil – fertility depends on biological factors, first of all on microorganisms." N.A. Krasilnikov, 1958.

Pierre Oyeaux, a renowned physicist wrote that each science experiences one or several heroic ages, marked by special investigating enthusiasm, when there is a hope to solve the "ever-lasting questions". This is followed by a classical period – maturity, when a circle of people dealing with a given science extends and specialization appears.

The science of chemical interaction of plants has obviously entered into a classical stage due to studies done under natural and artificial ecosystems. Allelopathy has attracted the attention of botanists, physicists, physiologists, microbiologists, soil scientists and other specialists. Microorganisms represent the most active part of any biogeocenosis and are characterized by enormous biochemical possibilities. According to Berestecki (1976), the allelopathic role of microorganisms was determined by the continuous synthesis of various physiologically active substances (antibiotics, phytotoxins, growth stimulators and inhibitors and others) and the conversions of physiologically active substances excreted by plants and microflora.

Krasilnikov (1958) showed that soil toxicity occurs due to accumulation of metabolites of soil microorganisms – toxins and other exometabolites and the microbes producing toxins are same in different types of soils.

However, the conversion and preservation of phytotoxic substances in soil as well as mechanism of their action on plants are little studied due to complex methods for their extraction from soil and their identification. Besides the specificity of phytotoxic metabolites of plants and microorganisms, mineralization products of plant residues and constant interactions between the various organic compounds and soil.

The incorporation of composts and plant residues in soil are not only sources of plant nutrients, but also of different physiologically active substances, which stimulate and inhibit the plants. This group of organic compounds included vitamins, auxins, soluble forms of humic acids and other biotic substances entering into the soil with plant residues, or produced by microflora (Krasilnikov, 1944; Mishustin, 1956; Bilay, 1961).

1. MICROFLORA

Krasilnikov (1958) found that biologically active substances were present in large amounts in soil. They enter into the soil as the products of microorganism metabolism and root excretions of plants. The quantity of physiologically active substances in the rhizosphere of crop plants was more than in deep layers of soil. Fertile soils rich in organic matter have more biotic substances than poor soils.

Soil toxicity was examined mainly in relation to depression of seed germination, inhibition of seedling growth and reduction in crop yields at harvest (Mirtshik, 1957; Tshunderova, 1964; Berestecki, 1971). Soil toxicity influenced the microbes association, in particular the separate groups of microflora – bacteria, fungi, algae (Krasilnikov, 1958). It had been noticed long ago, that soil had the ability to become free from pathogenic and some saprophytic microorganisms. The inhibitory effect of soil environment on saprophytic bacteria, fungi and some physiological groups of microorganisms was observed in sick soils. For example, application of nodule bacteria with nitragin into the clover sick soil, developed very little. There were similar data for *Nitrobacter* and it is a peculiar test-object for soil toxicity.

Krasilnikov (1958) reported that podzol soils were the most toxic, especially those, which were little cultivated. Toxicity of soils was determined by the survival of *Nitrobacter* and by germination of wheat seeds, as well as by the total count of microflora including microorganisms inhibitors. There were more toxic substances in forest soils than in ploughed soils, whereas in well cultivated soils, they were still less.

Mishustin (cited from Zaytshenko and Dachnovski, 1976) compared the biological activity and toxicity of soils at different levels. It is known that upper layers of soil have the highest biological activity and low toxicity to agronomically valuable microorganisms. Lower soil layers (40–50 and 50–70 cm) were characterized by low biological activity (estimated by decomposition of tissues and CO_2 production) and high toxicity in relation to wheat seedlings. Mishustin (1975) reported that soil toxicity can be determined by following parameters :

1. Chemical compounds of soil (pH, Al, Fe^{+2});
2. Toxic substances excreted by microorganisms;
3. Substances produced by microorganisms;
4. Bacteriophage activity;
5. Influence of high doses of mineral fertilizers;

Humus-ferric podzols have acidic pH of 4.0–4.5 and contain mobile ions of aluminium and ferrous oxide. Some humus components e.g. fulvic acids depressed the seedlings growth of crop plants. Rice straw contained 37% of coumaric and 22% of hydroxybenzoic acids, which can inhibit the seed germination. Phenolic compounds of sugar beet roots protected these plants from phytopatghogenic microorganisms but are toxic to other higher plants.

Similarly to many other biological properties of soil, the toxicity also changes with the season (Jukova, 1956) and the highest toxicity was noted in summer and autumn months. The toxicity decreases by late autumn and was lowest in the spring. It was due to (a) seasonal fluctuations in the number of microorganisms, (b) changes in toxins content due to washing out by rains in autumn and by melted snow in spring (c) partial inactivation by low temperatures and (d) sorption on soil particles. Toxic substances excreted by plants can be oxidized by some groups of microorganisms into simple compounds, which serve as source of nutritional substances. Under natural conditions, the koline accumulation proceeds according to the type of soils, nature of substance and outside environmental conditions. Some substances accumulate in significant amounts, while others were partially oxidized by microorganisms. In this way a rotation of physiologically active substances was created in biogeocenosis.

The toxic substances produced by some species of moulds, actinomycetes and bacteria play major role in soil sickness. At present, there are three main areas of research in this field:

1. Identification of different phytotoxic groups of microorganisms;
2. Species composition of different microorganisms;
3. Physico-chemical and biological properties of phytotoxic substances produced by microorganisms in crop fields and in natural conditions.

Berestecki (1976) found that turf podzol and peat-gley soils contained the largest population of phytotoxic microflora species. When plants are growing in monoculture, over 50% were phytotoxic fungal species, which were especially characteristic of perennial orchards. Among the bacteria, moulds, actinomycetes proactino-mycetes, micromonospores and others, there were always species, which were inhibitory under certain conditions.

1.1. Fungi

Among all microorganisms, fungi have the largest number of species producing toxic substances (Mirtshink and Stepanova, 1975). In turf-podzol soils the number of the main ecological groups of moulds, growing on yeast-agar ("sugar fungi"), Hutchinson medium (cellulolytic fungi) and on Chapeck medium (fungi with small food needs) amounted to tens and hundreds of thousands of fungal spores per 1 g of soil.

Toxins of saprophytic and phytopathogenic moulds, which have an influence on plants, are most important for soil. On this basis, complex interactions develop between microorganisms, soil and plant. Toxins are found in the soil (a) after its inoculation with cultures of fungi producing them, (b) natural soils containing organic substance in the form of plant fall and (c) in cultivated soils - at the occurrence of post-harvest residues (Krasilnikov and Garkina, 1976; Grodzinsky, 1974). According to the literature, the cultures of *Aspergillus terreus, A. clavatus, Penicillium patulum, P. nigricans* and *Trichoderma viride* accumulated the toxins in soil.

Mirtshhink (1976), reported that the fungus cultures *P. cyclopium, P. lapidosum* and *P. itrium* produced penicillic acid, patulin and citrinin. All three species were phytotoxically active and have wide antimicrobial spectrum, i.e. they were typical antibiotics. Phytotoxic activity (on germination of wheat seeds) of the given cultures was determined in field trials, soil after addition of 1% harvest residues.

Norstadt and Mc Calla (1971) showed production of patulin by the fungus *Penicillium urticae*, with the addition of wheat straw to soil. Under laboratory conditions, *P. urticae* synthesizes patulin on potato dextrose medium at 5–30°C and accumulated 2.7 mg/ml of patulin for 14 days at 25°C. Berestecki (1967) found the patulin production by *P. urticae* in sick soil under orchard fruit plants.

Mirchink (1976) found that podzol, turf podzol, grey-brown forest and brown taiga soils contain the largest quantity of toxic fungal species. Of all the microscopic fungal forms the most toxic in soil were of genus *Penicillium* (30%) and the less toxic were from group *Aspergillus* (12%). The genus *Penicillium* and its toxic forms commonly occurred in podzol and turf podzol soils, which are most toxic than those from peat soils. Berestecki and Sinicki (1973) noted that nearly 90% of the examined forms of *P. urticae* very phytotoxic to the germination of maize, pea and fiber flax. Hence, the addition of the most active forms of *P. urticae*, significantly decreased the germinating ability of the tested seeds.

1.1.1. Monoculture: The phenomenon of soil toxicity is strongly displayed under monoculture. Nadkernichny (1975) reported that the continuous cultivation of winter wheat promoted the growth of phytotoxic forms of microscopic fungi (10–12% in crop rotation, 42% in monoculture). The most frequently occurring species were: *P. islandicum, A. fumigatus, Aurobasidium pullulans, P. funiculosum* Thom.

Berestecki and Kalmykova (1974) and Kalmykova (1975) reported the increase in phytotoxic fungal forms under conditions of monoculture in winter wheat and maize on turf podzol and chernozem soils.

Lugauskas and co-workers (1976) grew clover and alfalfa, after various preceeding crops and in monoculture. The most typical associations of fungi microorganisms were in the rhizosphere and on the roots of red clover, especially in the monoculture: *Fusarium solani, F. gibbosum, F. oxysporum, P. purpurogenum, P. spinolosum* and others. These fungi excrete substances inhibitory to clover plants and are active destroyers of roots, hence, they can be the indicators of soil sickness.

The influence of crop plants on soil microflora and the influence of soil microflora on crop yield is clearly visible in monoculture. Bekker and Lebed (1966) investigated the dynamic of fungal antagonists in the rhizosphere. They showed that in flax monoculture (over 10 years) the total amount of fungal spores at all soil depths decreased by 4-5-folds. It is considered that fungal species with antibacterial activity are located in the deep soil layers (25-50 cm) and those with antifungal activity are in the upper humus layer of soil.

Romeyko and co-authors (1976) recommend the testing of phytotoxic activity of inhibitory microorganisms on seed germination of injured winter wheat seeds. They separated the inhibitory microorganisms from the root soil of winter wheat and maize grown in monoculture and in crop rotation in chernozem soil. Voynova and co-authors (1975) pointed out increased number of microorganisms participating in the nitrogen immobilization in soil, whereas the rotation of wheat, barley, sunflower and pea influenced the number of microorganism associations, which decomposed the organic substance.

Altimirskaya (1976) found that the winter wheat monoculture (4-5 years) on the leached black chernozem field, decreased the mineralization of organic substances and increased the nitrogen immobilization in the rhizosphere zone. Treatment of soil with 2 and 4% formalin and sowing of an intermediate maize crop increased the biological activity in soil.

1.1.2. Composition of Microflora groups: There is insignificant change in the total number and in composition of microflora groups in rhizosphere of crops grown in monoculture than in crop rotation. Burangulova and co-workes (1975) informed that continuous growth of winter rye, maize and spring wheat does not reflect on the total numbers of microorganisms. Growth of crops in crop rotation has the strongest influence on the numbers of cellulolytic and nitrifying microorganisms. On the leached chernozem soils of Bashkiria, the content of the main microorganism groups in different fields of grass varied depending on humidity, temperature, pH of soil and nature of crops in crop rotation.

In parabrown soil, the monoculture (for over 20 years) of cultivated crops (sugar beet, potatoes) strongly reduced the biological soil activity. However, periodically applied organic fertilizers in monocultures and introduction of crop rotation with cereals increased biological activity in soil. The numbers of microorganisms in the soil under monocultures differed insignificantly from their content in crop rotation, the biological activity under cereal monocultures (winter and spring wheat) was higher than that under intertilled crops (Beck 1975).

Sourovaya and Gorelik (1974) found that under monocultures of crop plants, changes occur in composition of the microflora groups, there are species of some spore-forming bacteria (*Bac. megatherium, B. cereus*) and moulds of the genera *Penicillium* and *Aspergillus*.

In this way, soil sickness observed under field in nature and under conditions of monoculture – is a complex phenomenon, which is determined by soil properties and toxic substances excreted by plants, derived from the plant residues and those produced by soil microorganisms. The accumulation of toxic substances can be activated by microflora using harvest residues as organic fertilizers, mulching favouring the development of fungal microorganisms of the genus *Penicillium* and *Aspergillus*, which produce patulin (clavatin, claviformin, clavacin), coumarin and organic acids (apple, citric, oxalic). Fungi participating in the mineralization of post-harvest residues can use humus substances as the source of nitrogen and to produce phenoloxidases and glucuronic acids, depressing the activity of nitrogen fixing and nitrifying bacteria, which results in the decrease of soil fertility. Toxic substances excreted by microorganisms suppress the seed germination and growth of higher plants. There are data showing that soil toxicity is displayed more strongly in relation to some groups microflora of higher plants than to higher plants. For instance, the inhibitory effect of fungal forms in relation to *Nitrobacter*, some species of spore-forming and cellulolytic bacteria.

Soil toxicity is most sensitively displayed in the root zone in plant rhizosphere. The share of plants and soil microflora in soil toxicity is little studied. There are no doubts that monoculture creates conditions for accumulation of the same kolines in the soil, which can determine the creation of an especial microflora association. Under conditions of monoculture, a change in composition of microflora groups in the plant rhizosphere, can inhibit the growth of useful microflora which produces, the vitamins, enzymes, organic acids and oxidizes the toxic substances. This explains the necessity of in depth comprehensive investigation of the role of microorganisms in allelopathic soil sickness in major field crops. The aim of our investigation was to learn the structure of microbial association of leached chernozem and grey forest soil, when growing agricultural plants in crop rotation and in monocultures, with the application of organic and mineral fertilizers in the control and under conditions of intensive monoculture (for over 45 years) and medium-intensive monoculture (15–20 years) taking into account stages of plant development. The seasonal changes in the number and composition of microflora groups in the rhizosphere of field crops was also determined to find the microorganism species indicating soil sickness.

For calculation of the total numbers of microorganisms in chernozem and grey forest soils in the rhizosphere and root zone of plants, we used medium of the soil extract (100 ml) with the addition of salts and agar-agar The total number and composition of separate microflora groups were counted by dish method inoculated with 0.5 ml of soil suspension from corresponding plant cultivation on parallel dishes. Bacteria chiefly using organic forms of nitrogen counted on meat-peptone agar (MPA), spore forming bacteria by inoculation of soil suspension (at 74°, expositure 10 min) on yeast-meat-peptone-agar medium (YMPA). As per Mishustin, the numbers of nitrogen fixing bacteria and actinomycetes were calculated on starch–ammonia agar (SAA), the numbers of moulds on acidified yeast-agar (YA) and in Chapeck medium. The number and activity of nitrifying and denitrifying microorganisms as well as the number of anaerobic nitrogen fixator of *Clostridium pasteurianum* were determined by the maximum cultivation method on the selective nutrition media. Aerobic nitrogen fixators – oligonitrophiles - were counted by dish method on the medium of Ashbi. Cellulolytic microorganisms (the total number, composition of separate groups and their activity) were studied on Hutchison's agar medium.

The main root mass is concentrated in the upper soil layer (0–25 cm). The biological importance of the root system is determined by its activity, i.e. the ability to absorb nutrients and produce carbonic acid, organic acids and release biologically active substances (koliens). These kolines are used by microorganisms as a source of nutrition, hence, they influence the microorganism accumulation in soil environment - the poorer the soil, the stronger the rhizosphere effect.

In soil samples of Mironovska, Drabovska and Poltavska experimental stations, the total numbers and group composition of microflora were counted from the rhizosphere of winter rye and wheat, maize, sugar beet and winter wheat. The soil samples with plants, taken from field conditions were soaked in distilled water and roots of separate plants were washed clean. After that, the roots with soil participles were cut in to small pieces and shaken up for 10 min on the rotor in previously weighed conical flask with 90 ml of sterile water. Soil suspension was applied on the appropriate dense and liquid nutrition media. The weight of soil of root zone was determined by the method of washing off soil participles on a previously weighed filter and by drying them up to a constant value.

Forster and Rovira (1976) also reported that the upper bark layer of the main and lateral wheat roots at the flowering stage were inhabited by microorganisms. Microflora of winter wheat rhizosphere was determined at the stages of tillering, stem extension, jointing and ripening of plants. The group composition of microorganisms in maize rhizosphere were stu- died at the stages of 2-4 leaf, silking, flowering milk yellow ripening.

Now major task of soil microbiologists is the comprehensive investigation of the ecology of microorganism populations entering into the composition of natural and cultural associations (Aristov-kaya, 1975; Zvyagintsev, 1975). The interrelations between the plants and soil microflora, which influence the activity of microorganisms and plant productivity, should be considered.

Evdokimova and Berednikova (1976) reported that the rhizosphere of fescue and timothy grass, grown on cultivated podzol soils of the Kola Peninsula, contained many toxin producing microorganisms. The relative content of stimulators in the roots zone is smaller than inhibitors and their stimulatory effects were weak on the plant growth. Antibiotic substances of the cultures of *Micobacterium globiformaanl, Pseudomonas fluorescens* were taken up by the plant through the root system and accumulate in the roots and in the lower part of the stem. Berestecki and co-workers (1976) observed an increase in the numbers of bacteria, bacilli and oligonitrophyles in turf, medium podzol soils in the crop rotation system in comparison with monocultures.

The first experimental studies on the influence of crop monoculture upon soil toxicity, total number and group composition of microflora were undertaken in a field trial at Drabovska station. In 1973-1974, we studied the seasonal dynamics of bacteria, microscopic fungi and actinomycetes, mainly using the organic forms of nitrogen.

The number of bacteria in the rhizosphere of winter wheat was higher in monoculture than crop rotation. The numbers of bacteria under intertilled crops were significantly higher (Table 49).

Table 49. Number of microflora in leached chernozem soil at Agricultural Research Station Drabovska

Crop	Cropping system	1973			1974		
		Bacteria (10^6 per g)	Fungi	Actinomycetes	Bacteria (10^6 per g)	Fungi	Actinomycetes
			(10^3 per g)			(10^3 per g)	
Winter wheat	Monoculture	6.7	129.0	171.4	7.9	16.5	21.5
	Crop rotation	13.3	75.1	98.4	2.2	41.3	15.8
Maize	Monoculture	7.4	49.8	52.3	2.9	29.7	13.8
	Crop rotation	11.2	50.8	75.6	2.1	27.3	19.5
Sugar beet	Monoculture	n.e.	n.e.	n.e.	3.3	23.9	39.1
	Crop rotation	n.e.	n.e.	n.e.	3.1	34.9	16.7
Fallow	Monoculture	8.0	76.6	43.7	0.6	11.7	19.7
	Crop rotation	23.9	37.4	54.8	2.9	27.2	12.9

The highest number of microscopic fungi in the rhizosphere of the winter wheat monoculture was noted in 1973. In the rhizosphere of maize significant fluctuations were observed in the number of mould spores at various vegetative stages. In 1974, there were more microscopic fungi in the rhizosphere of winter wheat grown in crop rotation and phytotoxic species were dominant. The highest number of moulds occurred in spring and in autumn, due to higher precipitation and incorporation of harvest-root plant residues.

Actinomycetes also produce antibiotics and phytotoxic substances into the environment. The maximum actinomycetes were found in the rhizosphere of winter wheat monoculture (171.4 th./g). Their population was less in the rhizosphere of intertilled crops and least in the fallow field. The number of actinomycetes in various treatments significantly differed in vegetation seasons, i.e. depended on hydrothermal conditions of the environment and the microorganism-dwelling zone with a mobile organic substance.

Table 50. Number and microflora group composition of leached chernozem in monoculture and crop rotation systems, Drabowo, 1974, 10^3 per g dry soil

Crop	Cropping system	Extract pH	Bacteria		Micro-organisms		
			Soil extract agar medium	Starch-ammonium agar medium	Cellu-lolytic	Nitri-fying	Deni-trifying
Winter wheat, rhizosphere	Monoculture	7.33	212.0	37.5	18.5	10.0	1000.0
	Crop rotation	7.85	174.0	74.0	9.0	100.0	1.0
Winter wheat, root zone	Monoculture	7.61	154.0	421.0	9.0	1.0	100.0
	Crop rotation	7.19	288.5	77.5	15.0	1.0	100.0
Maize, rhizosphere	Monoculture	7.99	45.0	267.5	12.0	1.0	10.0
	Crop rotation	7.73	204.0	48.0	7.5	100.0	100.0
Sugar beet, rhizosphere	Monoculture	7.30	148.0	35.5	6.0	100.0	100.0
	Crop rotation	7.00	158.5	25.0	9.0	100.0	10.0

In October 1974, we took soil samples including those from the rhizosphere and root zone of winter wheat with the aim to study the group and species composition of spore-forming bacteria. The rhizosphere of monoculture and crop rotation winter wheat contained approximately the same number of bacteria (Table 50). However, in the root zone of winter wheat grown in crop rotation system, they were 1.7-fold more. Monoculture of cultivated plants mainly contained those bacteria, which use the mineral nitrogen forms. A very sensitive group of microorganisms in relation to monoculture of field crops are nitrifying bacteria, their number was 1-2-fold less than in crop rotation. In the rhizosphere of monoculture winter wheat, cellulolytic fungi were predominant, whereas actinomycetes and fungi dominated in root crops. In crop rotation, a wider group composition was observed. Cellulolytic bacteria were constantly found, which is the index of well cultivated soils. Among spore-forming bacteria (Table 51), predominant were mainly *Bacillus mesentericus* and *Bac. megatherium*, constituting over 50% of the total numbers of this group.

Table 51. Species of bacteria occurring in leached chernozem, (Drabowo, 1974, 10^3 per g of dry soil)

Crop	Crop system	*Bacillus mesentericis*	*Bacillus megatherium*	*Bacillus cereus*	*Bacillus myco-ides*	Other species	Total
Winter wheat rhizosphere	Monoculture	20.0	22.0	4.0	15.0	21.0	82.0
	Crop rotation	19.0	12.0	0.5	12.0	7.5	51.0
Winter wheat root zone	Monoculture	12.0	24.0	2.0	7.5	4.0	49.5
	Crop rotation	17.0	14.5	2.0	5.0	13.5	51.0
Maize rhizosphere	Monoculture	13.0	11.0	1.5	4.0	11.5	41.0
	Crop rotation	18.0	13.0	0	4.0	15.0	50.0
Sugar beet rhizosphere	Monoculture	12.0	13.5	3.5	0	17.0	46.0
	Crop rotation	18.0	17.5	2.5	3.5	42.0	83.5

Studies on the influence of growth of monoculture field crops upon the total numbers and group composition of microflora were also conducted on dark-grey forest soils of Chartoryski experimental field (Zytomir district). Soil samples from the rhizosphere of winter wheat, sugar beet, maize and pea monocultures at different plant developmental stages and were also compared with their growth in crop rotation system (Table 52). The content of bacteria in the rhizosphere of monoculture of winter wheat and maize insignificantly differed from that of crop rotation. The largest amount of bacteria was found in the rhizosphere of sugar beet monoculture (92.8 mln/g of soil). Bacteria, mainly not spore-forming, dominated in the group composition of monoculture sowings of these field crops. In monocultures, the quantity of microscopic fungal spores during the vegetation phase was maximum (113 th./g) in the rhizosphere of pea and minimum (90 th./g) in the rhizosphere of sugar beet and maize. Plant growth in crop rotation changes only the content of embryos of microscopic fungal forms at different stages of plant development and even stimulates an increase in the numbers of a given group of microorganisms in the rhizosphere of winter wheat.

Table 52. Effect of monoculture of field crops on rhizosphere microflora, Chartorysky Research Field

Crop	Cropping system	1973			1974		
		Bacteria (10^6 per g)	Fungus spores (10^3 per g)	Actinomycetes (10^3 per g)	Bacteria (10^6 per g)	Fungus spores (10^3 per g)	Actinomycetes (10^3 per g)
Winter wheat	Monoculture	17.1	59.6	17.2	2.9	56.5	10.2
	Crop rotation	19.0	98.0	32.7	3.3	31.5	5.5
Sugar beet	Monoculture	92.8	93.6	2.0	3.5	27.2	11.5
	Crop rotation	33.8	67.3	55.0	3.3	19.0	14.0
Maize	Monoculture	32.3	90.6	19.0	1.7	10.0	8.1
	Crop rotation	20.4	74.6	69.0	1.9	36.0	3.4
Peas	Monoculture	23.7	113.0	58.2	6.2	25.5	7.2
	Crop rotation	20.0	80.0	17.6	2.4	20.0	7.4

A preliminary investigation of the total number and group composition of microflora of the rhizosphere of the main field crops grown in monoculture and in crop rotation, showed, that micro-organisms can serve as indicators of soil sickness. We found an increase in the number of spores of microscopic fungal forms in the rhizosphere of monoculture of winter wheat and maize. The content of microscopic moulds increases in the treatment with permanent fallow it has the lowest number of microorganisms, due to the decline of supplies of easily mineralized organic substances. There are some interrelationships between the cultivated plants and their accompanying microflora in monoculture, which influences the physiological activity of micro-organism groups and the plant productivity.

It is very difficult to analyze the group composition of microflora from the point of view of its participation in the creation of soil phytotoxicity. It may be suggested beforehand, that some species or groups of microorganisms producing phytotoxic antibiotics and having no tolerance to them, will disappear from population, though they appeared to be the cause of soil sickness. Besides that, a predominant development of any microorganism groups on sick soils is not a cause, but a consequence of soil sickness. The presence of these species is an indicator of soil sickness and for this reason, investigations on the dynamics of the number of microorganism groups can be useful.

Experimental data obtained in 1974 at Drabovska research station and from Chartoryski field trials offered a basis to select the most interesting treatments of field trials with the main agricultural crops. We did the studies on the seasonal dynamics of the total numbers and group composition of microflora in the rhizosphere of monoculture of field crops at these stations and in long-term field trials of Mironovsk Station (winter wheat and sugar beet in the field experiments being grown in monoculture since 1929 and maize since 1969). These crops were grown without fertilizers and with fertilizers (6 t of manure and $N_{10}P_{10}K_{15}$ annually) under winter wheat and inter-tilled crops. At Drabovsk and Chartoryski fields, soil samples were taken from rhizosphere of winter wheat, maize, sugar beet and pea grown in monoculture for 15-20 years, the influence of organic and mineral fertilizers applied to these crops was also taken into account.

Under conditions of monoculture of winter wheat and particularly maize and sugar beet at Mironovsk station, some acidity of soil (pH 6.35 – 6.51) was observed. Mould fungi separated by us from the rhizosphere of monoculture intensively transformed the simple carbohydrates to organic acids and caused soil acidity. At Drabovsk experimental station, the influence on the rhizosphere of monoculture winter wheat was neutral. A small decrease of pH value towards acidity was observed in the rhizosphere of maize and sugar beet grown in crop rotation system. However, a significant change in the pH value from spring to autumn was noted in all treatments of the field trial. Medium alkalinization was observed in spring and some acidification in September. This is explained by supply of plant-harvest residues and by intensive biochemical activity of microscopic fungal forms in autumn period.

In this way, under the influence of maize, sugar beet and winter wheat monoculture, acidification of the environment can be observed, which increases the soil toxicity and influences the interrelations between separate groups of microorganisms.

2. MICROFLORA OF WINTER WHEAT MONOCULTURE RHIZOSPHERE

Winter wheat is main food in USSR. However, soil sickness begins to occur due to the influence of monoculture of agricultural crops, leading to a decrease in plant productivity and to an increase in soil toxicity.

2.1. Bacteria

A comparative study of the number of bacteria which chiefly use the organic forms of nitrogen, has shown that their number in the rhizosphere of monoculture winter wheat is 2-7-fold higher than the wheat grown in crop rotation system (Table 53). The application of organic and mineral fertilizers in winter wheat do not change the number of heterotrophic bacteria. A stimulatory action of manure and mineral fertilizers has been observed on the number of a given group of microorganisms in the rhizosphere of winter wheat grown in crop rotation. The number of rhizosphere microorganisms insignificantly differed in vegetation seasons but depended on the development stage of winter wheat and on hydrothermal conditions of the environment. The largest number of bacteria in the rhizosphere of winter wheat was found at the stages of tillering and stem extension and the lowest number at the ripening phase. This is explained by plant excretions of carbohydrates, organic acids and amino acids into the soil as sources of energy and nutrition of soil microorganisms during intensive growth of plants. The decrease in the number of microflora in the rhizosphere of plants grown in monoculture can increase the soil toxicity at definite phases of plant development and particularly in dry season. Therefore, in winter wheat soil sickness can occur most actively in autumn, during shoot appearance, when mineralization of plant-harvest residues proceeds intensively and the number of microflora increases, and in summer during the period of intensive plant growth.

The destruction of post-harvest root residues is most intensive in monoculture and crop rotation, because it is the most mobile and easily mineralizing source of organic substances. A gradual increase in the number of given group of bacteria from 1.7 to 4.5 mln/g of soil was observed in the rhizosphere of monoculture winter wheat during crop seasons. In crop rotation system, these indices were 1.5–3-fold higher in the unfertilized soil. The application of manure and mineral fertilizers increased the number of bacteria in the rhizosphere of winter wheat grown in monoculture.

Out of all groups of microorganisms, the most interesting were bacteria using primarily humus soil substances (calculated on soil agar medium), they are called oligocarbophylles. We examined the seasonal dynamics of the given group of microflora. Their number constituted from several hundreds to tens of millions per 1 g of soil.

At Drabovsk station, monoculture of winter wheat did not decrease the total numbers of oligocarbophylles in the rhizosphere in comparison to crop rotation. Bacteria in the rhizosphere of this culture were 1.5 – 3-fold lower than from Mironovsk field trial crop rotation. This peculiarity was observed in the vegetation seasons and at separate stages of development (Table 53). This indicates an increase of soil toxicity under winter wheat monoculture, because oligocarbophylles were depressed by the dominating group of bacteria, which use primarily simple organic substances. The decrease in the number of prevailing groups of bacteria favours the increase in number of microorganism acidifying the humus substances of soil. The given group of bacteria excretes physiologically active substances (vitamins, enzymes, auxins) into the soil which increase the productivity of growing plants.

Table 53. Bacteria number in leached chernozem in winter wheat growing in monoculture and crop rotation systems, Research Station, Mironovsk (10^6 per g of dry soil)

Cropping system	Fertilization	1975			1976			1977		
		May 15	June 26	July 27	June 18	July 24	Sept. 6	June 15	July 27	Sept. 15
Bacteria on meat-pepton agar medium										
Monoculture	Unfertilized	1.55	0.65	0.54	0.12	13.9	3.10	1.29	0.83	1.31
	Fertilized	1.30	0.94	0.55	0.13	3.5	3.05	0.64	1.02	1.31
Crop rotation	Unfertilized	0.53	1.92	0.30	0.16	0.8	0.52	0.83	1.19	4.69
	Fertilized	2.96	2.26	0.28	0.07	4.1	3.02	0.47	1.25	10.11
Bacteria on starch-ammonia agar medium										
Monoculture	Unfertilized	7.43	2.21	1.21	0.25	19.9	17.3	4.59	5.34	3.72
	Fertilized	4.34	1.87	1.21	0.19	29.5	28.80	9.26	21.18	5.44
Crop rotation	Unfertilized	2.58	2.02	0.45	0.34	28.4	24.30	2.77	15.15	6.65
	Fertilized	2.24	2.06	1.35	1.15	18.9	23.80	2.49	5.54	16.65
Bacteria on soil extract agar medium										
Monoculture	Unfertilized	1.04	1.27	1.69	0.45	16.5	0.38	3.12	30.37	2.08
	Fertilized	1.03	1.88	1.35	0.21	30.3	0.16	9.88	114.97	1.58
Crop rotation	Unfertilized	3.51	4.12	3.14	0.22	19.1	1.40	5.20	42.04	2.01
	Fertilized	2.90	2.48	1.18	0.39	20.6	0.87	1.21	208.11	17.75

Table 54. Effect of winter wheat monoculture on group composition of microflora, Research Station, Mironovsk (10^6 per g of dry soil)

Cropping system	Fertilization	1975			1976			1977		
		May 15	June 26	July 27	June 18	July 24	Sept. 6	June 15	July 27	Sept. 15
Microscopic fungi										
Monoculture	Unfertilized	e409.5	14.0	40.0	3.6	87.1	12.4	117.16	90.0	1.18
	Fertilized	129.9	27.3	21.0	2.4	207.3	19.8	89.85	41.7	5.95
Crop rotation	Unfertilized	85.6	21.4	12.0	10.4	65.1	20.1	41.00	50.0	9.50
	Fertilized	14.0	51.0	31.0	4.6	225.3	7.9	25.87	67.4	14.46
Actinomycetes										
Monoculture	Unfertilized	144.3	105.0	120.0	18.2	45.6	45.0	278.40	69.2	224.90
	Fertilized	259.0	88.2	210.0	12.2	621.0	180.0	151.70	359.2	166.70
Crop rotation	Unfertilized	90.9	153.0	40.0	5.8	100.3	73.0	146.00	128.0	237.60
	Fertilized	18.4	106.0	178.0	5.8	251.0	67.0	105.80	260.2	144.60
Spore-forming bacteria										
Monoculture	Unfertilized	81.0	31.0	41.0	14.6	213.8	80.4	176.30	272.30	90.70
	Fertilized	61.0	39.0	33.0	12.2	159.8	66.9	238.00	208.60	157.20
Crop rotation	Unfertilized	53.0	31.0	28.0	20.9	26.9	35.8	330.30	854.30	138.90
	Fertilized	33.0	35.0	48.0	17.4	162.7	41.7	169.30	80.4	140.90

2.2 Microscopic fungi

Bacteria were predominant in the group composition of microflora. Among microscopic fungi, the toxic species were found most frequently. For this reason, study of the mould dynamics under field grown crops in monoculture and in crop rotation system is important, to determine the role of microorganisms in soil sickness. The largest amount of spores of microscopic fungi was observed in the rhizosphere of winter wheat monoculture (Table 54). The content of spores of microscopic fungi in the rhizosphere of winter wheat varied significantly in the crop seasons and at various stages of plant development. An increase in the number of spores of microscopic fungi was also observed at the stages of tillering and stem extension. The lowest content of mould fungi occurred in September (1.18–12.4 th./g of soil). In the rhizosphere of winter wheat monoculture at Drabovsk station, the number of mould fungi was from 19.0 to 299.0 th./g of soil, seasonal changes being clearly observed in their number. The lowest mould content was in summer (May), while the highest content in September-October.

Investigation of biological properties of microscopic fungi separated from the rhizosphere of winter wheat showed that in the vegetation season of 1976, over 50% of them had phytotoxic properties. Some forms of microscopic fungi from the genera *Penicillium, Aspergillus,* developed normally under laboratory conditions on a liquid medium of Chapeck and produce abundant spores. The acidifying of nutrition medium (pH 4.65–6.41) and in inhibited the increase in number cress rootlets from 12 to 24.1% and of wheat seedlings from 15 to 46.2%.

2.3 Actinomycetes

Under the influence of winter wheat monoculture, the number of actinomycetes in the rhizosphere increased 1.4–2.5-fold during the vegetation period, in comparison to this wheat grown in crop rotation system (Table 54). The application of manure and mineral fertilizes under monoculture of winter wheat increase the number of actinomycetes. For example, the average content of actinomycetes in the rhizosphere of wheat monoculture was 185.0–225 th./g of soil than 98.0-170 th/g in wheat sown in crop rotation. In actinomycetes, there are some species with phytotoxic properties and capable to inhibit the some bacterial associations. Intensive inhibitory zones, primarily of the genus *Pseudomonas* were observed on Petri dishes.

Seasonal dynamics of the number of actinomycetes in the rhizosphere of winter wheat monoculture is described by the gradual transitions from spring to autumn, the absolute indices of vegetation periods significantly differ. The actinomycetes number increased mainly at the beginning of intensive vegetation of plants and the entry of fresh organic substances. In the rhizosphere of winter wheat grown in crop rotation system, some increase occurred in the number of actinomycetes at the jointing stage of plants. In soil with fertilizers, actinomycetes were significantly more numerous, but seasonal changes in their numbers were identical to those considered above. This group of microorganisms is biologically important producer of phytotoxic substances, owing to its physiological and biochemical interrelations with other microflora species.

2.4 Spore forming bacteria

An essential group of soil microorganism associations is constituted by spore-forming bacteria, occurring in the form of spores (bacillus, bacilli). The total number of spore-forming bacteria in the winter wheat monoculture in the vegetation periods was 51-100 th./g, it was 1.5-4-fold higher than in crop rotation (Table 54). This phenomenon should be explained by peculiarities of plant growing in monoculture. Groups of spore-forming bacteria, poor in species composition, have the most favourable conditions for their existence in the rhizosphere of monoculture crops. Growing of winter wheat in the soil with mineral fertilizers significantly reduced the numbers of spore-forming bacteria, because of inhibitory effect of other groups of microorganisms, for instance, phytotoxic species of mould fungi, actinomycetes and no spore-forming bacteria. Seasonal changes in the spore-forming bacteria were most clearly displayed in the vegetation periods of 1976 and 1977, which insignificantly differed in hydrothermal conditions. A low number of spore-forming bacteria was noted in spring, in the beginning of plant vegetation, which later increased from tens to hundreds thousands in summer and declined in September. This displays an essential difference in the dynamics of spore-forming bacteria in comparison to microscopic fungal forms and actinomycetes, which are the main producers of phytotoxic substances. In studies of soil sickness under field crops, the species composition of spore-forming bacteria was very distinct. *Bacillus mesentericus, Bac. elaterium, Bac. subtilis* were predominant in the rhizosphere of winter wheat grown in monoculture. In crop rotation, the species composition of spore-forming bacteria is very diverse, i.e. besides the above species, *Bac. agglomeratum, Bac. cereus* and others occurred sporadically.

We observed the inhibitory interrelations the between different species of spore-forming bacteria. Antagonistic interrelations were seen under conditions of winter wheat monocultures between particular species of spore-forming bacteria (Fig. 23). In Petri dishes, we observed the depression of *Bacillus mycoides* by toxins of *Bac. subtilis*. In our many cultures of spore-forming bacteria, *Bac. mycoides* was the most sensitive species, which can serve as a test-object in the studies of soil sickness.

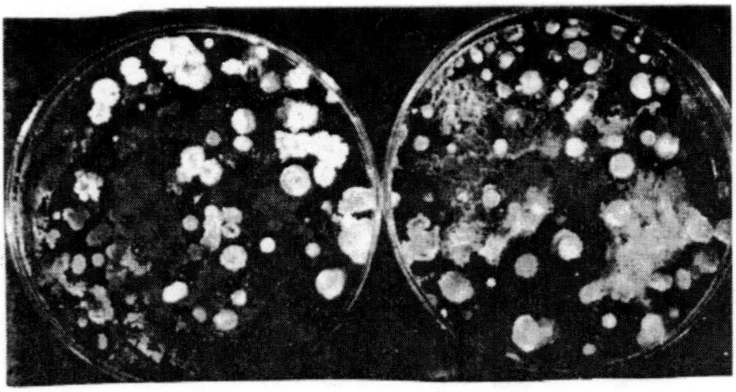

Fig. 23. Species composition of spore-forming bacteria of winter wheat rhizosphere from monoculture.

2.5 Oligonitrophiles and Oligocarbophiles

Soil environment is constantly inhabited by microorganisms getting nitrogen and carbon nutrient from solutions with very low amounts of these compounds. These are oligonitrophiles and oligocarbophiles microorganisms constituting a group of soil microflora completing mineralization of organic compounds. Zavarzin (1972) called this group "dispersing microflora". This can be a joint name for a complex of microorganisms using low contents of monomers formed by other microorganisms, since they are able to develop at very low content of mineral nitrogen in the environment. This property provides the possibility of their existence under conditions unfavourable to other species of microorganisms. The rhizosphere of winter wheat monoculture was found to have from several hundreds to tens of millions of oligonitrophylles per 1 g of soil, i.e. the content of the given group of microorganisms decreases insignificantly under the influence of monoculture in comparison to crop rotation (Table 55). The application of manure and fertilizers under monoculture of winter wheat sowing, essentially stimulated the increase in oligonitrophiles number. The number of oligonitrophylles in the rhizosphere of winter wheat monoculture grown under conditions of Drabovsk station was from 2.4 to 50.7 mln/g of soil, which is less than their number in crop rotation. However, the given numbers of oligonitropylles are comparable and this indicates that certain associations of microorganisms can exist under conditions of monoculture. According to data of Mironovsk field trial, the seasonal dynamics of oligonitrophylles showed an increase in their number in the period of intensive plant vegetation. At Drabovsk Station, this increase was observed in September and can be explained to some degree by the entry of fresh organic substances, poor in nitrogen.

Table 55. Dynamics of oligonitrophile number in rhizosphere of winter wheat, Research Station, Mironovsk (10^6 per g of dry soil)

Cropping system	Fertilization	1975			1976			1977		
		May 15	June 26	Aug. 27	June 18	Aug. 26	Sept. 6	June 15	July 27	Sept. 15
Monoculture	Unfertilized	2.77	2.96	1.46	0.09	0.49	0.23	44.93	44.93	0.29
	Fertilized	2.05	2.13	2.24	0.15	2.94	2.72	66.62	66.62	0.38
Crop rotation	Unfertilized	1.07	3.99	0.94	0.24	0.49	0.44	9.63	7.89	0.13
	Fertilized	4.42	4.39	2.54	0.24	0.43	-	0.06	108.20	0.10

The number of oligonitrophylles exceeds 2-4 fold, the content of common saprophytes assimilating mainly organic forms of nitrogen. This increase in the number of oligonitrophiles strongly fluctuates at different stages of plant development. Bacteria and actinomycetes constantly dominated among the oligonitrophiles. The role of oligonitrophiles microorganisms is connected mainly with their ability to fix small amounts of molecular nitrogen. Klevenska (1974) reported that oligonitrophilous microorganisms can bind 1.7-4.8 mg of nitrogen per 1 g of carbon containing substance.

2.6 Cellulose microorganisms

Cellulose is one of the main components of plant residues and hence, it has significant influence on the formation of some biochemical properties of soil. From the point of view of the studies on soil sickness, it is necessary to know the number and group composition of cellulolytic microorganisms. Cellulolytic bacteria separated by Jukova (1959) from the soils of the Kola Peninsula were inhibited by antibiotic substances of microscopic fungi from the genera *Penicillium* and *Aspergillus*.

Growing of winter wheat in monoculture, determined some inhibition in the total number of cellulose microorganisms in comparison with crop rotation. In the rhizosphere of winter wheat monoculture during the vegetation season, their number varied from 44.1 to 112.0 th./g (Table 56). Organic and mineral fertilizers insignificantly stimulated the increase in the total number of given microorganisms. At Drabovsk Station, the number of cellulose microorganisms in the rhizosphere of winter wheat monoculture were 19.2 – 195 th./g, whereas their number under conditions of crop rotation was less (22.7 – 100 th./g). Seasonal changes in the number of cellulose microorganisms are clearly observed in the treatments. The number and activity of cellulolytic microbes increase chiefly in May and September, which was connected with the entry of post-harvest residues and with the improvement of hydrothermal conditions of the environment. Cellulolytic microorganisms are used as indicator of soil fertility. It is considered that fungi dominates in the soils of low fertility.

Table 56. Number of cellulolytic microorganisms in rhizosphere of winter wheat, Research Station, Mironovsk (10^6 per g of dry soil)

Cropping system	Fertilization	1975			1976			1977		
		May 15	June 26	Aug. 27	June 18	Aug. 26	Sept. 6	June 15	July 27	Sept. 15
Monoculture	Unfertilized	80.0	33.0	19.0	18.2	16.9	298.8	4.21	44.90	85.2
	Fertilized	61.5	30.5	12.5	69.3	13.3	277.8	3.08	67.01	84.5
Crop rotation	Unfertilized	54.0	31.5	14.9	63.8	5.0	518.1	9.63	7.89	62.9
	Fertilized	56.0	46.0	19.0	13.9	12.5	59.6	3.06	108.26	75.9

With the improvement in physical and chemical properties of soil, bacteria of the genus *Cellvibrio* begin to appear, which indicates that the soil is provided with mineral nitrogen, and some species of *Cytophaga* appear in fertile high-humus soils (Peat-swampy soils). The group composition of cellulolytic microorganisms of the rhizosphere in winter wheat monoculture included mainly microscopic fungi (12.0–91.7%) and bacteria (8.3-19.1%). Seasonal changes were observed in the group composition of cellulolytic microorganisms. Microscopic fungi dominated at the beginning of plant vegetation, the number of bacteria was less and later this relation between individual groups stabilized, whereas in September actinomycetes prevailed (31.5-80.6%) and the content of microscopic fungal forms decreased (Table 57). The comparison of the dynamics of the group composition of cellulolytic microorganisms in treatments and at the stages of plant development showed that soil toxicity increases in June (fungi dominate) and in September (phytotoxic species of actinomycetes predominate). On this basis, it can be concluded that the group composition of cellulolytic microorganisms is an indicator of soil sickness.

In this manner, monoculture of winter wheat contributes to soil acidification particularly in autumn. The number of bacteria, actinomycetes and spores of fungal flora increased in the rhizosphere of winter wheat monoculture, many of the separated forms of microorganisms have phytotoxic properties. The highest content of microorganisms using primarily humus substances was noted in the rhizosphere of wheat monoculture. Under the influence of monoculture, the content of oligonitrophylles insignificantly decreases in comparison to winter wheat grown in crop rotation, however, the given association of microorganisms plays an important role in biological fixing of atmospheric nitrogen.

Seasonal studies of the dynamics of the number of major groups, showed the stimulation of bacterioflora development at the stage of tillering and stem extension but further their number decreased, especially in the period of ripening. This is explained by a decrease in the excretion activity of plants and by accumulation of phytotoxic substances produced by plants and some species of microorganisms. The dynamics of microscopic fungal forms was clearly observed during the phases of plant development. At the beginning of vegetation, the number of fungal flora increased to hundreds of thousands, then stabilized and decreased in August. However, in the rhizosphere of winter wheat monoculture, an increase in fungal flora was observed at crop ripening. In other groups of microorganisms in the seasonal dynamics, there occurred a gradual increase in their number at the beginning of plant vegetation and in autumn.

3. MICROFLORA OF WINTER RYE MONOCULTURE

Winter rye had been grown in monoculture at Poltavska experimental station since 1885. This is the third field in the world (after Rohamsted experimental station and the Experimental field of Timiriazev Agricultural Academy). During this period, the humus content in chernozem decreased by 0.35%. In the first 10 years, the yield of winter rye decreased constantly and there after it stabilized at 12.0 dt/ha. Though winter rye belong to plants with a high allelopathic activity, thus it suppresses various weeds (Grodzinsky, 1973), it is simultaneously autotolerant, which permits to grow this crop quite successfully in monocultures, especially with the application of organic and mineral fertilizers.

We examined above microorganism groups of soil rhizosphere, depending on the conditions of winter wheat growing. These groups of microorganisms can be differently developed and presented by specific genera and species. Vinogradski (1951) has introduced a concept of zymogenous and autochthonous microflora of soils. They include microorganisms decomposing the fresh plant residues. A zymogenous group includes microorganisms which oxidize the humus substances in soil. They constitute the most numerous group of microflora. On this basis, it is advisable to compare the microorganism groups in the rhizosphere of winter rye grown in monoculture and crop rotation.

Table 58 presents data showing seasonal changes in the number of bacteria using primarily organic forms of nitrogen and the numbers of autochthonous microflora. Rhizosphere of winter rye grown in monoculture, insignificantly differed in the content of microflora from the rhizosphere of winter rye in crop rotation system. However, at the jointing stage and at the beginning of ripening, in the rhizosphere of winter rye grown in crop rotation system, microorganisms population was 1.5-2.5-fold more especially heterotrophic bacteria.

A positive influence of organic and mineral fertilizers was observed also at the stages of jointing and beginning of ripening. Bacteria whose numbers ranged from 0.5 to 7.5 mln/g dominated in the group composition of microflora. Spore-forming bacteria were less (Table 59) i.e. tens and hundreds of thousands per 1 g. The growth of winter rye favours the some increase in the number of spore-forming bacteria. The species composition of spore-forming bacteria was poor, main species were *Bacillus megatherium* and *Bac. mesentericus* (rhizosphere of winter rye monoculture) and some species had inhibitory activity. Fig. 24 shows inhibitory zones of *Bac. mycides* and other species of spore-forming bacterium *Bac. mesentericus*.

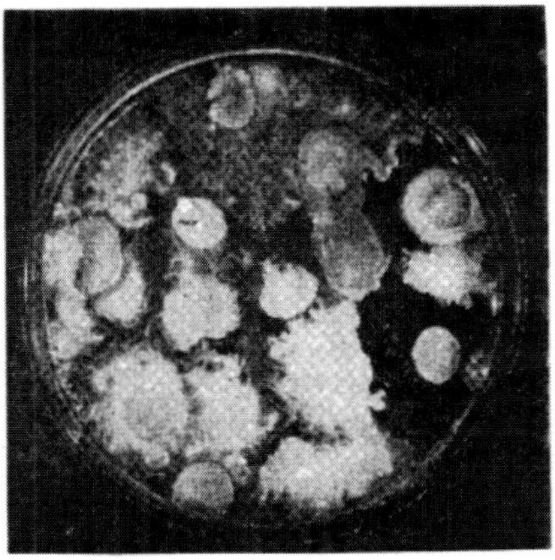

Fig. 24. Spore-forming bacteria of winter rye rhizosphere from monoculture.

Microscopic fungi were counted on the Chapeck medium and yeast-agar. Their population was maximum in the rhizosphere of winter rye monoculture with the application of fertilizers (55.4-215 th./g) and were from the genera *Penicillium, Trichoderma* and *Fusarium* having phytotoxic properties. It should be mentioned that these data were obtained by the method of soil suspension culture on dense nutrition media and for this reason; they constitute only a small part of the total number of microscopic fungal forms. (Mirchink, 1972) quantitatively calculated the microscopic fungi on membrane filters. Calculations of fungal flora biomass in chernozem and podzol soils show that their biomass corresponds to the bacterial biomass or a little less. The species composition of soil moulds was identified on solidified agar nutrient media. Representatives of the genus

Penicillium separated from the rhizosphere of winter rye monoculture had phytotoxic properties. They inhibited germination of radish and growth of seedling rootlets of cress in laboratory experiment. Table 60 presents the dynamics of the number of oligonitrophylles belonging to autochthonous group and the total number of cellulolytic microorganisms.

In the rhizosphere of winter rye monoculture, the number of oligonitrophylles was 0.74–5.70 mln/g and in the crop rotation was 1.74–9.55 mln/g. The number of cellulolytic microorganisms was several tens of thousands in monocultures, which insignificantly differed from the crop rotation systems. It was observed that the group composition of cellulolytic microorganisms changes with seasons (Table 61). In the spring, cellulolytic fungi (50%) and actinomycetes (50%) dominated in the rhizosphere of winter rye monoculture. Actinomycetes (70%) were dominant in the rhizosphere of crop rotation winter rye, whereas, fungi with cellulolytic activity constituted 30%. Later, at the jointing stage and at the beginning of ripening, cellulolytic fungi (72-80%) were dominant in the rhizosphere and some bacteria (8.5-28.0%) occurred. Biochemical activity of cellulolytic microorganisms in associations of microflora provides oligonitrophiles with mobile sources of carbon-containing substances : organic acids and alcohols. Oligonitrophilous micro-organisms are able to fix atmospheric nitrogen independently and in symbiosis with other groups of microflora.

On the whole, the microflora of the rhizosphere of winter rye grown in monoculture insignificantly differed from that grown in crop rotation. Bacteria dominated in the group composition. Among microscopic fungi, the dominating species were phytotoxic. The species composition of spore-forming bacteria is poor and some frequently occurring species had an inhibitory activity. Fungi and actinomycetes dominated in the group composition of cellulolytic microorganisms and these groups of microflora had phytotoxic properties.

4. MICROFLORA INTERTILLED MONOCULTURE CROPS

The main crops in the Ukraine are sugar beet, maize and potato. The influence of crop monocultures upon soil microflora (total number and group composition) was studied in long-term field trials at Mironovsk, Drabovsk and Chartoryski Experimental Station. At Mironovsk, sugar beet had been grown in monoculture since 1929 and maize since 1959. Treatments were compared also on the background of organic and mineral fertilizers (6 t of manure and $N_{10}P_{10}K_{15}$). At Drabovsk and Chartoryski, maize and sugar beet had been grown in monoculture for 15-20 years and soil also was fertilized. Seasonal investigations on the dynamics of pH in leached chernozem showed, that in the rhizosphere of maize, especially in crop rotation with sugar beet, the soil was slightly acidified than monoculture. The change of pH was observed in these treatments from spring to autumn. In spring, alkalization of the environment was observed. In summer and in September–October, leached chernozem soils have high buffer abilities of soil solution.

4.1 Bacteria

Out of all groups of microorganisms, the most interesting are bacteria using primarily the humus soil substances which are also called oligocarbophiles. Their population in the rhizosphere of root and tuber crops varied from several hundreds to thousands tens of millions per 1 g of soil (Table 62). Maize grown in monoculture, did not reduce the total number of oligocarbophiles in the rhizosphere in comparison to crop rotation. However, in the period of intensive growth, the number of microorganisms in the rhizosphere of plants grown in crop rotation was 1.5-3-fold higher than in monoculture. The application of organic and mineral fertilizers in monocultures of root and tuber crops reduced the total number of oligocarbophiles. Root and tuber crops grown in monoculture increased the mineralization of organic soil substances, due to favourable influence on the number of oligocarbophiles and the decreased inhibitory influence on other groups of microorganisms. The number of oligocarbophiles, increases in summer and decreases in May and September due to seasonal dynamics.

A comparative investigation of the number of bacteria, using primarily organic forms of nitrogen, has shown that values of absolute indices of the given group are significantly lower in comparison with oligocarbophiles. The content of bacteria in the rhizosphere of maize monoculture varied from 0.12 to 114.0 mln/g of soil (Table 63). Fluctuations in the number of bacteria in the rhizosphere of maize grown in crop rotation were measured approximately in the same limits. Thus monocultures of maize insignificantly differed from the crop rotation in the number of bacteria, because root excretions of plants create favourable ecological conditions for given group of microorganisms. This phenomenon was observed in the treatments with the growth of monoculture sugar beet. The application of organic and mineral fertilizers in monocultures of root and tuber crops mainly had a positive effect on rhizosphere bacteria. Seasonal changes in the number of rhizosphere bacteria in monoculture of maize and sugar beet depended on the stages of crop development and they were closely interrelated with the dynamics of hydrothermal conditions of the environment.

In the rhizosphere of maize, an increase in number of bacteria was observed at the stages of eighth-ninth leaf and at panicle initiation stage the fertilized treatments insignificantly differed by indices of seasonal changes of microorganisms. The lowest content of bacteria was found at the stage of stem extension and at the beginning of plant vegetation. On this basis, it can be affirmed that soil toxicity under monocultures of the root and tuber crops was most significantly displayed during the period of intensive plant growth. Microflora developing in the root zone has positive influence on the plant nutrition, especially at the beginning of vegetation. Soil microorganisms can partially decrease the soil sickness under monoculture, because they use the phytotoxic substances produced by plants. The microbe associations consisting of the microorganisms capable of metabolizing the root excretions of plants and the oxidizing toxic substances, can act the in monocultures. It is necessary to study the dominating and especially soil sickness indicator species viz., spore-forming bacteria, microscopic fungi and actinomycetes.

4.2 Microscopic fungi

The content of microscopic fungi in the rhizosphere of monoculture maize in the vegetative seasons and at stages of plant development varied from 20.0 to 96.0 th./g of soil (Table 64). The content of mould fungi in the rhizosphere of maize grown in crop rotation system slightly exceeded the above values.

Monocultures and crop rotation of sugar beet insignificantly differed in the number of spores of microscopic fungi. However, at separate stages of plant development, the number of moulds was 1.5-3-fold more in the rhizosphere of monoculture. In contrast to the dynamics of the number of microscopic fungi in the rhizosphere of these crops at Drabovsk Station, under conditions of long-term Mironovsk experiment, the mould number increased during the summer. Among the microscopic fungi, phytotoxic species of the genus *Penicillium* (10-12% toxic forms), *Aspergillus* (14-19%), *Fusarium* (4-5%) etc., were found. Predominance of phytotoxic species of microscopic fungi at certain stages of plant development partially determines the seasonal change in soil toxicity observed under monoculture. Ten cultures of microscopic fungi isolated from the sugar beet monoculture had phytotoxic properties. These cultures acidified nutrition medium to pH 2.71-5.56 and in bioassays with cress showed inhibition from 19 to 80%. Among the separated microscopic fungi, form No. 4, acidified the nutrition medium to pH 2.0-2.5 and caused 100% inhibition of seed germination of radish and inhibited the growth of cress rootlets and coleoptiles of winter wheat.

Phytotoxic forms of microscopic fungi, which acidified nutrition medium to pH 4.51-6.00, were also separated from the rhizosphere of sugar beet grown in crop rotation with organic and mineral fertilizers; some of them maintained neutral pH of the medium. Some species stimulated the growth of cress seedlings in culture medium. Some species of mould fungi inhibited the root growth of cress from 17 to 46%. Thus among the phytotoxic forms of microscopic fungi, separated from the rhizosphere of root and tuber crops, the largest number belongs to the genera *Penicillium* and *Aspergillus*.

Under the influence of maize monoculture, the number of actinomycetes in the rhizosphere during the vegetative phase decreased by 1.4-2.5-folds in comparison to crop rotation (Table 64). For this reason, monoculture growth of plants impoverishes the soil environment by slight decrease in the number of actinomycetes. The application of manure and mineral fertilizers in monoculture of root crops increased the number of actinomycetes. This phenomenon was clearly observed in the monoculture of sugar beet, where the number of actinomycetes in the rhizosphere increased by 1.5-3-folds. A stimulatory effect of the applied fertilizers was observed in maize monocultures in 1977, because the number of actinomycetes increased from 70.3 to 307.0 th./g of soil.

Table 57. Group composition (%) of cellulolytic microorganisms in rhizosphere of winter wheat, 1977

Cropping system	Fertilization	June 15			July 27			August 15		
		Bacteria	Fungi	Actino-mycetes	Bacteria	Fungi	Actino-mycetes	Bacteria	Fungi	Actino-mycetes
Monoculture	Unfertilized	19.1	80.9	0	8.3	91.7	0	8.3	11.2	80.5
	Fertilized	58.8	41.2	0	10.8	89.2	0	7.2	12.6	80.2
Crop rotation	Unfertilized	22.5	77.5	0	12.0	78.0	10	32.0	13.2	54.8
	Fertilized	13.8	86.2	0	22.8	77.2	0	23.8	26.9	49.3

Table 58. Dynamics of bacteria number in soil samples, Agricultural Research Station, Poltavska 1977, (10^6 per g of dry soil)

Treatment	pH	Bacteria on meat-pepton agar medium			Bacteria on extract soil agar medium		
		April 26	May 26	July 5	April 26	May 26	July 6
Fallow in crop rotation	7.30	0.48	9.71	0.90	2.60	1.06	6.14
Winter rye in crop rotation	7.21	0.60	45.30	2.39	2.13	3.20	7.06
Winter rye monoculture, unfertilized	7.20	0.63	7.57	1.39	5.06	2.64	4.40
Winter rye monoculture, fertilized	7.21	0.65	3.55	0.73	2.91	3.40	7.40

Table 59. Number of spore-forming bacteria and microscopic fungi in soil samples, Agricultural Research Station, Poltavska 1977, (10^6 per g of dry soil)

Tereatment	Spore forming bacteria			Microscopic fungi		
	April 26	May 26	July 5	April 26	May 26	July 5
Fallow in crop rotation	101.5	175.4	305.7	38.9	103.0*/117.0**	24.0*/27.4**
Winter rye in crop rotation	255.3	122.3	147.2	49.1	41.3*/97.8**	21.8*/33.3**
Winter rye monoculture, unfertilized	137.7	65.9	467.7	43.0	40.4*/92.5**	55.4*/54.0**
Winter rye monoculture, fertilized	83.2	100.7	547.6	55.4	64.8*/25.10**	24.0*/43.3**

* = yeast agar medium, **Chapeck medium

Table 60. Number of oligonitrophylles and cellulolytic microorganisms, Agricultural Research Station, Poltavska 1977

Treatment	Oligonitrophylles, (10^6 per g)			Cellulolytic microorganisms, (10^3 per g)		
	April 26	May 26	July 5	April 26	May 26	July 5
Fallow in crop rotation	1.58	1.50	8.06	28.31	27.3	53.8
Winter rye in crop rotation	1.74	3.14	9.55	31.60	65.2	42.6
Winter rye monoculture, unfertilized	5.70	5.13	0.74	77.60	49.7	38.1
Winter rye monoculture, fertilized	3.49	3.54	7.96	62.40	44.7	53.6

Table 61. Group composition of cellulolytic microorganisms in soil samples, Agricultural Research Station, Poltavska 1977 (%)

Treatment	April 26			May 26			July 5		
	Bacteria	Fungi	Actino-mycetes	Bacteria	Fungi	Actino-mycetes	Bacteria	Fungi	Actino-mycetes
Fallow	0	100.0	0	37.5	62.5	0	90.9	6.0	3.1
Winter rye in crop rotation	0	33.4	66.6	28.0	72.0	0	14.9	85.1	0
Winter rye monoculture, control	0	52.6	47.4	16.3	83.7	0	8.5	89.4	2.1
Winter rye monoculture fertilized	0	48.2	51.8	17.5	82.5	0	0	100.0	0

Table 62. Effect of monoculture and crop rotation systems of crop plants on number of bacteria using humic substances of soil, in long-term experiment, Mironovsk (10^6 per gram of dry soil)

Crop plant	Crop system	Fertilization	1975			1976			1977		
			May 15	June 26	Aug. 27	May 18	July 24	Sept.6	June 15	July 27	Sept.15
Sugar beet	Monoculture	Unfertilized	5.41	4.47	2.90	0.64	7.7	10.5	5.31	31.7	28.70
		Fertilized	3.91	3.07	0.83	0.14	4.8	2.4	1.47	85.9	3.40
	Crop rotation	Unfertilized	4.91	3.42	0.82	0.17	10.8	1.3	1.99	29.4	29.90
		Fertilized	2.10	1.62	0.71	0.07	17.5	1.1	16.50	82.6	2.03
Maize	Monoculture	Unfertilized	7.30	8.71	1.24	0.29	17.1	0.42	2.14	63.31	1.64
		Fertilized	3.19	2.31	0.90	0.54	21.1	0.33	4.44	266.4	1.73
	Crop rotation	Unfertilized	5.10	4.72	0.99	0.35	19.4	0.57	2.29	91.7	20.2
		Fertilized	3.31	2.42	0.47	0.09	6.2	-	2.84	92.4	17.1

Table 63. Dynamics of bacteria number using organic nitrogen in leached chernozem, Research Station, Mironovsk (10^6 per g of dry soil)

Crop	Cropping system	Fertilization	1975			1976			1977		
			May 15	June 26	Aug. 27	May 18	July 24	Sept.6	June 15	July 27	Sept.15
Sugar beet	Monoculture	Unfertilized	0.39	0.69	0.56	0.45	4.70	5.60	2.04	2.04	2.61
		Fertilized	0.44	2.39	0.34	0.26	3.60	2.70	7.17	7.17	6.35
	Crop rotation	Unfertilized	0.95	1.58	0.35	0.01	0.53	0.70	1.08	1.08	1.86
		Fertilized	1.05	0.77	0.24	0.55	0.47	0.48	5.40	5.40	4.45
Maize	Monoculture	Unfertilized	1.39	1.29	0.35	0.12	4.10	2.70	14.00	14.00	1.34
		Fertilized	0.39	2.58	0.27	0.33	18.60	2.70	1.36	1.36	1.43
	Crop rotation	Unfertilized	1.53	2.67	0.20	0.15	8.40	15.60	1.07	1.07	1.18
		Fertilized	2.77	3.03	0.22	0.25	18.70	4.10	1.15	1.15	1.33

Table 64. Effect of crop monoculture on group composition of microflora, Long-term Experiment at Mironovsk (10^3per g of dry soil)

Crop	Fertilization	1975			1976			1977		
		May 15	June 26	July 27	June 18	July 24	Sept. 6	June 15	July 27	Sept. 15
		Microscopic fungi (10^3 per g of dry soil)								
Maize	Monoculture	46.5	37.4	20.0	14.3	96.7	27.1	29.1	30.4	5.76
	Crop rotation	159.6	62.2	19.0	13.3	100.1	44.7	26.5	31.7	30.20
Sugar beet	Monoculture	44.5	48.1	25.0	13.0	101.7	15.8	25.3	40.0	13.60
	Crop rotation	44.5	17.0	35.0	4.8	65.8	25.2	21.0	90.4	10.70
		Actinomycetes (10^3 per g of dry soil)								
Maize	Monoculture	8.0	30.0	90.0	9.0	58.5	57.0	97.0	70.3	126.70
	Crop rotation	13.7	207.0	155.0	18.1	47.6	45.0	48.3	190.2	192.20
Sugar beet	Monoculture	9.9	63.0	90.0	11.8	79.1	22.5	96.7	80.1	216.20
	Crop rotation	19.0	42.0	120.0	5.4	112.0	22.4	70.8	46.2	202.10
		Spore-forming bacteria (10^3 per g of dry soil)								
Maize	Monoculture	43.0	64.0	28.0	14.3	42.4	4.7	271.4	190.5	81.70
	Crop rotation	92.0	63.0	49.0	41.1	84.5	17.6	322.5	144.8	165.10
Sugar beet	Monoculture	19.0	178.0	36.0	28.4	16.9	3.4	107.0	169.4	134.00
	Crop rotation	29.0	70.0	45.0	44.1	44.6	25.2	276.3	158.3	213.40

Table 65. Oligonitrophilic and cellulolytic microorganisms in sugar beet and maize rhizosphere, long-term experiment at Research Station, Mironovsk

Crop	Fertilization	1975			1976			1977		
		May 15	June 26	July 27	June 18	July 24	Sept. 6	June 15	July 27	Sept. 15
		Oligonitrophiles, (10^6 per g of dry weight)								
Maize	Monoculture	2.43	8.54	0.72	0.64	1.46	0.75	2.58	13.12	0.49
	Crop rotation	0.91	4.47	0.51	0.65	1.25	0.32	3.86/	24.41	0.16
Sugar beet	Monoculture	0.61	6.85	2.27	0.47	0.45	1.39	1.55	18.83	0.01
	Crop rotation	0.94	4.21	0.94	0.19	0.63	1.33	1.45	42.73	3.17
		Cellulolytic microorganisms, (10^6 per g of dry weight)								
Maize	Monoculture	48.00	36.50	4.24	8.40	2.30	254.40	2.58	13.12	47.20
	Crop rotation	48.60	21.50	22.00	56.90	3.60	144.70	3.86	24.41	57.10
Sugar beet	Monoculture	51.00	27.00	10.19	41.50	5.10	896.60	1.55	18.83	22.00
	Crop rotation	92.00	41.50	37.00	53.60	13.10	159.10	1.45	42.73	61.00

Seasonal dynamics of actinomycetes in the rhizosphere of root and tuber crops has distinct regularities in each vegetation period. The lowest content of actinomycetes 8.0-0.9 th./g was in spring, at the 2-4 leaves stage, later it increases to several tens of thousands in summer months and was highest in autumn. Among the actinomycetes there are species inhibitory to the genus *Bacterium*. The role of actinomycetes in soil sickness consists in producing toxic substances inhibiting the development of agronomically important physiological groups of microorganisms.

4.3. Spore forming bacteria

We have attributed the spore-forming bacteria to a group of microorganisms which reveal the phenomenon of soil sickness. In fact, there were less spore-forming bacteria in the rhizosphere of monoculture of maize and sugar beet in comparison to their population in crop rotation (Table 64). The number of spore-forming bacteria varied in the rhizosphere of maize monoculture during the stages of plant development from several tens to hundreds of thousands per 1 g of soil. The maximum spore-forming bacteria (82 to 272 th./g of soil) occurred in 1977; in maize rhizosphere. Seasonal changes have less affected the number of spore-forming bacteria, because the association of microorganisms react sensitively to the changes of hydrothermal conditions in the environment. Some increase in the number of spore-forming bacteria was observed in June and in August. The application of manure and mineral fertilizers significantly inhibited the development of microorganisms of this group, because the dominating bacterial flora and microscopic fungi inhibited the number of spore-forming bacteria. In this relation, the given association of microorganisms is a sensitive indicator of change in ecological conditions of microflora dwelling and the display of soil toxicity. The most common species in the composition of spore-forming bacteria were *Bacillus megatherium, Bac. mesentericus, Bac. cereus, Bac. subtilis* and other species. Inhibition in development on agar plates of *Bac. mycoides, Bac. subtilis* by the metabolites of *Bac. mesentericus* was observed. The spore-forming bacteria separated from the sick soils have higher biochemical activity than its other species (Figs. 25, 26).

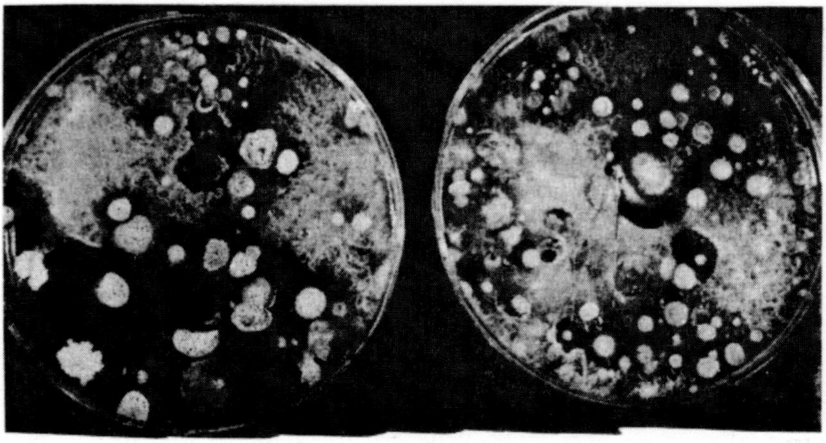

Fig. 25. Spore-forming bacteria of rhizosphere from maize monoculture.

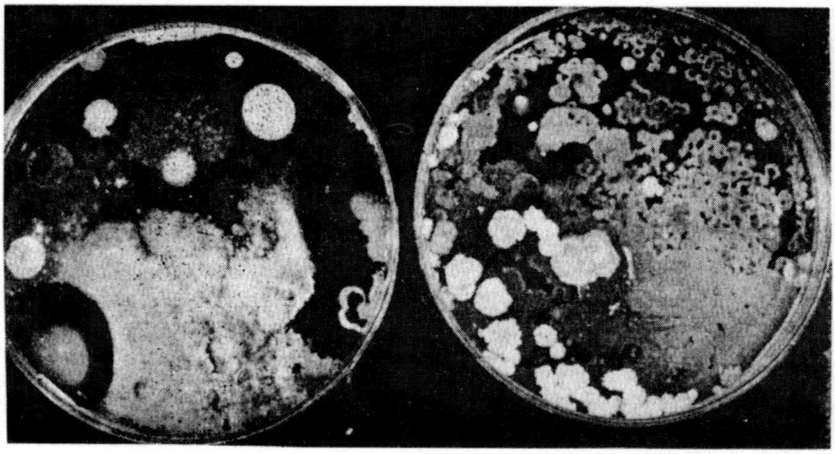

Fig. 26. Spore forming bacteria of rhizosphere from sugar beet monoculture.

4.4. Oligonitrophiles

Oligonitrophilous microorganisms constitute the most numerous group of autochthonous microflora developing on poor nutrition media and using primarily small amounts of monomers of soil environment. Oligonitrophiles have the ability to develop on a very small content of nutrients in soil solution, and due to this property they can survive under unfavourable ecological conditions. Table 65 shows the dynamics of the number of oligonitrophilous microorganisms in rhizosphere of crops. Monocultures of maize and sugar beet insignificantly differ in the number of oligonitrophiles than their growth in crop rotation system. However at different stages of plant development, the crop rotation stimulated the number of oligonitrophiles to several tens of millions per 1 g of soil.

In the season dynamics of oligonitrophiles, an increase in the number was noted during the period of intensive plant growth. The lowest number of oligonitrophiles was observed in August and September due to decrease in the excreting activity of crops cultures. Bacteria dominated in the group composition of oligonitrophiles, whereas actinomycetes occurred episodically. Actinomycetes are able to excrete antibiotics having a lytic, bactericidal or bacteriostatic activity, which influence the other groups of microorganisms. Klevenskaya (1974) reported that the intensity of inhibitory effects of actinomycetes depends on the antibiotics released by them into the environment. *Actinomyces roseoflavus, Act. olivaleoviridis, Act. violceus* producing flavomicin and violarin have the strongest influence on oligonitrophiles.

4.5. Cellulolytic microbes

Associations of microorganisms consisting of cellulolytic bacteria, actinomycetes, microscopic fungi and oligonitrophiles occur under natural conditions. These associations function due to the significant supplies of cellulose coming into soil form crop residues.

It is known that most of oligonitrophiles have no cellulolytic activity, but they use the intermediate products of cellulose decomposition. Under anaerobic conditions, for instance, when growing rice, hydrolysis of cellulose comes with accumulation of organic acids (acetic and butyric acids) which are used by oligonitrophiles.

The importance of cellulolytic microorganisms in the associations mentioned above is in providing oligonitrophiles with mobile carbon-containing substances. In these associations, participating in hydrolytic decomposition of cellulose under low content of nitrogen in the medium and intensive nitrogen fixation by oligonitrophiles was be observed. Therefore, oligonitrophiles can provide cellulolytic microorganisms with easily mineralized nitrogen. In the microflora associations developing on the basis of cellulose hydrolysis, there is higher stability of pH than in pure cultures of cellulolytic microorganisms.

Cellulolytic microorganisms are peculiar indicators of ecological conditions of soil environment. In acid and podzol soils, microscopic fungi dominates in the group composition, while in fertile, chernozem, peat–bog soil, cellulolytic bacteria of the genus *Cellvibrio* were dominant. The mould fungi developing in podzol soils may inhibit cellulolytic bacteria.

The rhizosphere of monoculture contained very low amount of cellulolytic microorganisms (4.24-48.0 th./g of soil) and weak stimulation was observed in fertilized soil (Table 65). Maize monoculture was less toxic in relation to cellulolytic bacteria than monoculture of sugar beet, during the phase of intensive plant growth. The influence of soil sickness on the number of cellulolytic bacteria was strongly displayed at the stages of germination and 2-4 leaf of sugar beet grown in monoculture. In 1975 - a dry year, toxic substances accumulated in upper layers of soils, hence the dynamics of cellulolytic bacteria showed a decrease in their number from spring to autumn. A seasonal change in soil toxicity was also peculiar to maize monoculture. Inhibition of cellulolytic microorganisms under monoculture was also observed in the vegetation season of 1977, because the number of the given group of microflora varied within 1.55-22.0 th./g of soil. However in maize rhizosphere, an inhibitory influence of soil sickness on cellulose bacteria decreased from the beginning of plant vegetation until milk-ripe stage of corn. Thus seasonal changes in soil toxicity were determined by (a) koline accumulation in the upper layers of soil, (b) an increase in phytotoxic activity of microorganism groups and (c) hydrothermal conditions of the environment.

The group composition of cellulolytic microorganisms of monoculture rhizosphere consisted primarily of microscopic fungi (80.5%) and a small amount of bacteria (19.5%). Microscopic fungi (93.7%) dominated in the rhizosphere of sugar beet and Cellulolytic bacteria (3.1%) occurred sporadically. In crop rotation, the number of microscopic fungal forms decreases and the content of cellulolytic bacteria increases to 17-33%. The crop rotation also intensifies the biochemical activity of cellulolytic microorganisms.

Seasonal changes in total number of cellulolytic microorganisms were clearly observed in their group composition. At the beginning of crop vegetation, at the stage of 2-4 leaves, cellulolytic fungi were dominant and bacteria were less. This correlation between the main groups of cellulolytic microorganisms was continued in subsequent phases of plant development. However, in September, cellulolytic actinomycetes prevailed in the rhizosphere of these cultures (50%), the content of bacteria increased (30%) and the number of microscopic fungi decreased (12-14%). The comparison of the dynamics of group composition of cellulolytic microorganisms in cultures and developmental stages showed that the given microflora association can serve as a bioassay, when studying soil sickness.

5. MONOCULTURE/CROP ROTATION

In this way, monoculture of cultivated plants increases the mineralization of organic soil substance and favours the activity of rhizosphere oligocarbophiles. Monocultures of maize and sugar beet using primarily the organic forms of nitrogen, differed insignificantly in the content of bacteria. This is because the field crops release root exudates containing organic acids, amino acids and biologically active substances, which have positive effect on the development of given group of micro-organisms. In monocultures, spore-forming bacteria did not prevail in the group composition of rhizosphere microflora; spores of microscopic fungi numbered from some tens to hundreds of thousands per 1 g of soil and the number of phytotoxic species increased.

Study of species composition of phytotoxic fungi has shown that the largest number of species with phytotoxic activity belongs to the genera *Penicillium, Aspergillus* and *Fusarium*. Domination of phytotoxic forms of fungi at certain stages of plant development (germination, 2-4 leaves) induces certain degree of soil toxicity. The phytotoxic forms of fungi acidified the nutrition media to pH 2.0-2.5 by producing citric, oxalic and apple acids and induced 100% inhibition in bioassays. Species composition of spore-forming bacteria, the total number and group composition of cellulolytic are quite sensitive indicators of soil sickness.

In monocultures of cultivated crops, the number of cellulolytic bacteria decreases. In sick soils, the number of cellulolytic bacteria decreases, whereas the content of fungal flora increases. Seasonal changes in the number of bacteria under maize and sugar beet depended mainly on the developmental stage of plants and the hydrothermal conditions of environment. Soil toxicity under monoculture of root and tuber crops was maximum during the period of sprouting and intensive growth of plants. In monoculture, the microflora developing in the root zone can partially decrease in the soil sickness, some species of microorganisms can decompoose the phytotoxic substances.

A comparative investigation of the total number and group composition of rhizosphere microflora of the main crops grown in monoculture and in crop rotation system, have shown that microorganisms can be the indicators of soil sickness. In growth of monoculture winter wheat at Mironovsk station, the development of oligotrophic bacteria was stimulated. In sugar beet and maize monoculture, mineralization of organic substance of soil and the number of microscopic fungi and bacteria was increased. The application of organic and mineral fertilizers in monocultures of field crops increased the biochemical activity of the main groups of microflora.

The total number and separate groups of microorganisms changes depending on the phase of plant development and hydrothermal conditions of the environment. The rhizosphere of winter wheat, the number of bacteria increased at the stages of tillering and stem extension, and then their content reduced during the crop ripening. In monocultures of root and tuber crops, soil toxicity was maximum during the periods of germination (sprouting) and intensive plant vegetation. Monocultures of maize and sugar beet differed insignificantly in the total number of bacteria.

The plant excretions and post-harvest residues can induce soil sickness through promoting certain associations of microorganisms. In monoculture, growth of winter wheat and rye, sugar beet and maize insignificantly changes the number of the main groups of microorganisms than crop rotation system. Associations of oligocarbophiles and oligonitrophiles (from 2.4 to 50.7 mln/g of soil) occurs in the rhizosphere of crop monocultures.

In the rhizosphere of winter wheat monoculture, the spore-forming bacteria numbered from 24 to 535 th./g of soil. The species composition of spore-forming bacteria is an important indicator of soil sickness, because in monoculture the spectrum of species is limited due to the inhibition by toxins of some species by others.

In monoculture of field crops, the phytotoxic species of microscopic fungi are dominant. They cause 30 to 100% inhibition in seed germination of radish and depressed the rootlets of cress and wheat seedlings. Some of these species intensively acidified the culture medium to pH 2.0-2.5 by producing citric, apple and oxalic acids and some phenolic substances.

Cellulose is one of the main components of crop harvest residues, it influenced the formation of biological soil properties and the domination of a certain group of cellulolytic microorganisms, which indicate the soil sickness.

Chapter 8

DETERMINATION OF ALLELOPATHIC SOIL TOXICITY

A.M. Grodzinsky

Department of Allelopathy,
Central Botanical Garden, 1, Timirvzskaya Street, Kiev, Ukraine

"To know this, we must ask this question the plant itself....."
K.A. Timiriazev, Selected works (1949)

The presentation of allelopathy as a cycle of physiologically active substances in biogeocenosis, suggests studies on the accumulation of kolines and their conversion in soil, without which the mechanisms of chemical plant interaction cannot be discovered. The allelopathic soil sickness can be separated from other forms of soil sickness, by a direct definition of active substances accumulating in soil.

The extraction of physiologically active substances from soil for their identification and quantitative evaluation is most important. Soil is the strongest adsorbent, which adsorbs many substances including kolines (Grümmer, 1957; Pogrebniak, 1968). The soil does not prevent allelopathy and is an active participant in chemical interactions of plants. Generally, when plant extracts are passed through soil layer, their activity in bioassays did not decrease, but increased (Moroz *et al.*, 1971; Moroz, 1972a). This can be explained by exchange adsorption of physiologically active substances (Grodzinsky, 1978), which results in release of earlier adsorbed compounds, whose toxicity can be higher than kolines of the initial extract.

Owing to enormous sorption ability, the soil adsorbs and retains physiologically active substances (kolines) released from root or leaf excretions or decomposing plant residues. The soil does not permit them to evaporate or percolite into deep soil layers and in this way accumulate kolines in certain ecosystem zones, to play their allelopathic role. Hence, the determination of kolines content in these or other points of ecosystem is major objective of allelopathy evaluation.

Kolines having diverse chemical nature do not remain unchanged in the soil. Starting from physico-chemical properties of soil, it can be assumed that koline interaction with soil proceeds in the following way:

- chemical conversions of active compounds - hydrolyses, oxidation, restoration, neutralization, nitrification decarboxylation, amination, polymerization etc.;

↓

- adsorption on clay minerals; humus complex;

↓

- biological adsorption by soil flora and fauna with corresponding transformations;

↓

- remaining in a free state.

The adsorption on clay minerals and organic colloids is of different nature viz., (a) mechanical, (b) molecule sticking in pores and cracks, (c) intermolecular attraction, (d) electrostatic attraction, (e) formation of hydrogen bridges, (f) covalent bonds and (g) inclusion into crystalline structure.

Bear and Thomas (1964, 1965) found the mechanism of interactions between the volatile physiologically active substances and soil. Clay minerals of sedimentary rocks adsorbed volatile substances, and a specific complex called petrichor was formed (see Chapter 5). It is supposed that strongly attracted molecules in the petrichor zone are accumulated on the surface of clay mycelia. Various reactions can occur between them and even spontaneous fixation of atmospheric nitrogen also occurs, which is included in the composition of nitro-compounds. It seems to us that this phenomenon reminds a peculiar catalysis similar to adsorption catalysis, which is used in chemical industry. The most diverse compounds – from large chelato-similar molecules to elementary sulphur – enter into the composition of petrichor. It is probably important for soil sickness that nitrophenols formed in petrichor are similar to herbicides in their nature and cause a very strong suppression of plants.

Organic substances in soil undergo mineralization processes, i.e. decomposition to CO_2, H_2O and to simple mineral salts or humification - polymerization and formation of comparatively stable large humus molecules, humins and fulvic acids. Though the both processes are represented as opposite, but they produce the most stable and low-entropic products. In molecule mixture due to natural selection, unstable high-energetic ones are gradually removed from the medium, while the most inert ones remain. In all intermediate stages of these two processes, there are low molecular mobile substances, which are physiologically active and determine the allelopathic soil effects.

There are aerobic and anaerobic soil chemical processes related to periodical excessive moisture and over drying of soil, when oxygen is forced out or returns again. The process of humification began on aerobic or anaerobic pathway is stopped by mineralization, hydrolysis, depolarisation, etc. In the end, a larger part of organic substance "is burnt" to mineral ingredients, simultaneously the low-molecular substances appear, which play the role of kolines.

The mobility of volatile or water soluble kolines is comparatively low in soil. The molecules which are not retained by this or other mode in soil are washed out or evaporated. Only those kolines act on plants, which can be taken up by roots or which can penetrate into a plant or can act on plants indirectly, through regulation of accompanying microflora, or, due to change in physical soil properties.

1. EXTRACTION

The kolines easily extracted by water are less in soil and all other remaining kolines are strongly adsorbed and are in the complex with inactive compounds, from which they can liberate only after enzymatic or acid hydrolysis. The plants are able to uptake mineral elements from the adsorbed state: they indicate soluble ability of roots in relation to tri-basic calcium phosphate. The roots under sterile conditions are able to hydrolyze the starch and other complex organic compounds with the help of extracellular enzymes. Thus plants are capable to uptake kolines from an adsorbed state as well as from their complex, little mobile compounds. It may be possible that symbiotic microflora enhancing the plant absorbing ability participates in the mobilization of kolines.

Therefore, researchers have a difficult task to find the method for koline extraction, which could explain the soluble and absorption ability of roots in relation to organic substances. As the kolines have diverse chemical nature, their extraction from soil by one solvent is not possible. Besides, their extraction can be accompanied by larger or smaller changes, such as oxidation and restoration, hydrolytic degradation, etc. depending on the methods of extraction. Kolines can react with solvent, between each other, with air oxygen and so on. Therefore, different methods can lead to highly variable results. This question is complicated by out studies on very different soils.

Extraction of water soluble kolines by water, which is used most frequently is less efficient. Even at very narrow ratios of soil to water (1:1.5; 1:1 and 1:0.5), the obtained amounts of extracted organic compounds were very low. The next difficulties in substances extraction is that the organic substances in soil are bound with 2- and 3-valent cations and probably with hydrogen bridges of carboxyl or hydroxyl groups of phenolic or aliphatic molecules and also amino groups (to a lesser degree) with clay and silica minerals. Hence, many different extraclants are used for separation of organic substances from soil (Tinsley and Salam, 1961; Dubach and Mehta, 1963).

1.1. Solvent

As per the chemical and physical properties, solvents can be divided into several groups: (1) acids; (2) strong alkali; (3) weak alkali; (4) complex agents; (5) organic solvents.

A preliminary treatment of soil with acids was used for desorption of humus substances, which are fixed on tetrazol oxides, free silicic acid, or on clay minerals and for removal of carbonates and exchanged bases. Soil treatment with hydrochloric acid or with a mixture of hydrochloric and hydrofluoric acids, accompanied by watering out until a complete removal of calcium leads to an increase in solubility in alkali and to a more complete extraction (Stevenson, 1957; Bremner and Harada, 1959; Choudri and Dubach, *et al.* 1963; Oposner, 1966). However, this method is not suitable for our purposes, since mobile acid organic products presented by kolines go away in acid waters and at neutralization they precipitate again as calcium ions.

Schnitzer and Wright (1956) extracting podzol soils with the diluted mineral acids (HF, HCl, or their mixture) at 0.5% concentration, found that the acids extract only 2% of the total content of carbon in the layer A_0 and up to 96% from layer B. Higher concentrations of acids are less effective. Besides that, there are correlations between the content of carbon in substances extracted by 0.5% HCl and the quantity of iron and aluminium. The maximum of C, Fe and Al contents were found in extracts from B layer. Ponomareva (1947), also found that large quantities of fulvic acid fractions are dissolved by cold diluted hydrochloric acid of podzol layer B in Tundra soils, but very small quantities are extracted from chernozem and grey forest soil.

Parsons and Tinsley (1960) dissolved 23% of organic substance from lime meadow soil by washing out with formic acid and precipitation with di-izopropylene ether. However, other changes occurred such as hydrolyses of carbohydrates, proteins and other polymers.

Generally water solutions of hydroxyperoxides of alkaline metals are used, as they provide the largest amount of organic substances from extraction. However, there are some objections against the use of alkali.

(a) Solution of silicic acid involving contamination of humus substances.

(b) Solution of undecomposed residues and fresh plant tissues including hemicellulose and lignin.

(c) Strong chemical changes, such as polymer hydrolyses to amino acids, sugars or condensation reactions between nitric compounds and phenols and probably aldehydes, which should lead to the formation of artificial humic acids.

(d) Extracted substances in the alkali solutions easily oxidize by air oxygen.

Polar solvents (alkali and acids) have high dielectric constants and it increase the solubility of humus substance by breaking hydrogen bonds, fix metallic cations and immobilize them and are the most effective extractants causing the lowest chemical changes (Whitehead and Tinsley, 1964). Besides, they can extract more than, what is available to plants, i.e. they can distort the actual picture.

Humus substances are also extracted from the soil by sodium salts and different acids. The most effective were NaOH – 96%; sodium phosphate – 94; sodium pyrophosphate – 92%; sodium fluoride – 83%. In this way, the amount of extracted substances increases with the increase in pH of the extractant; neutral solutions of pyrophosphate and fluoride extract more than sulphate chloride or orthophosphate at the same pH values. The output of extraction for all solvents increases in the increase of temperature.

To avoid undesired binding together of organic substances with mineral soil components, chelating agents are used, such as sodium ethylene diaminotetraacetylene (EDTA), ethyl acetate etc.. With the use of Na_2-(EDTA) (Flaig and co-workers, 1975) extracted 97% of total carbon from podzol soil layer B at pH 4.2 and only trace amounts from the A_0 layer. The short coming in this method is the biological activity of EDTA and the impossibility to determine the content of nitrogen and carbon in the obtained product. Such extractants as ethyl acetate, 8-hydrooxychinolin and their metal complexes can be easily removed from the extract by lipophilic solvents. Preliminary treatment of soil with acid or chelate also has an advantage, that humus substances are soluble in such organic solutions as acetone, diamethyl foramide, pyridine, dioxan, tetrahydrofuran, particularly, if they are mixed with water.

1.2. Extraction methods

Thus, all methods of organic substance extraction from soil can be divided into two groups: (a) destruction of soil particles by hard treatment with alkali or acids and (b) delicate extraction of substances adsorbed by colloids, clay or silicate minerals, without destruction of molecular structures of soil particles. We are interested in the second group of methods, which can model better sorbing ability of roots. Thus for koline extraction from soil we can recommend very weak acids, which permit to model the dissolving function of roots. Thus, for determination of phosphorus available to plants, 1% lactic acid is sometimes recommended. Taking into account that phenolic compounds are an important group of kolines, good results can be obtained using alkalized alcohol (Wang *et al.*, 1967; Hennequin and Juste, 1967; Careva and Semenova, 1971). Acidified ether can be also used (Terentiev *et al.*, 1969) as well as acetone, pyrophosphate of potassium, alkali and acid solutions (Guenzi and Mc Calla, 1966 b). The application of organic solvents makes possible the extraction of free compounds from soil, because humic acids are not extracted by them and adsorbed substances are not concerned.

Undoubtedly it would be very interesting to study the exchange reactions in soil and development of a suitable method. Obviously the mechanism of molecule absorption by plants from adsorbed state consists of the fact that root excretions participate in exchange with previously adsorbed kolines taking their places. The experiments of Morozov and Leshenko support the above possibility.

In separate cases, it could be useful to extract kolines by successive washing out by a series of solvents distributed according to a decreasing or increasing lipophylity. The problem is that the accumulation of salts in some soils and the accumulation of fatty-like substances in other soils can prevent from extracting kolines completely. In such approach to this problem, we gradually remove one layer after another so, that all mobile substances will be removed.

Another fractionated method, useful for the extraction of indolo-derivatives is shaking of soil suspension with acidified water, alkalization of water fraction by sodium carbonate and a repeated extraction by ether is another fractionating method, useful for the extraction of indolo-derivatives. Acidified ether fraction will contain IAA, whereas alkalized - indoloacetonitril. Water residue, which is usually disregarded, can contain physiologically active substances. For the extraction of volatile kolines from the soil, the method of fractional distillation should be tested in nitrogen presence or with water steam, as it is done in obtaining essential oils. Distillated substances should be collected in cooled condenser.

Let us consider some methods used to extract the active substances. To obtain free phenol acids, Hennequin and Juste (1967) took 50 g sample of dry soil sieved through 2 mm screen, shook for 2 h with 75 ml 0.2 N NaOH. The suspension was centrifuged for 5 min at 4000 rpm. The decanted liquid was acidified by a concentrated HCl to pH 1.0 and left over night for sedimentation of humus fraction, which was separated by centrifugation.

Phenolic acids were extracted from 50 ml of obtained filtrate (fraction of fulvic acids) by shaking with 50 ml of ether. The ether fraction was evaporated, dry residue was dissolved in 0.75–1.0 ml of ethanol and was plotted on the chromatogram and separated by 2% acetic acid. It was displayed by p-dinitroaniline.

Wang and co-workers (1967) used another method to study phenolic acids of soil. Approximately 100 g soil was poured with 95% ethanol immediately, after taking from the field to suppress microbial activity. Not decomposed organic residues were removed and pH of the suspension was increased to 11 with 2 N NaOH. The suspension was left until the next morning and, if necessary, pH was increased again to 11. It was shaken for 1 h and centrifuged. The volume of supernatant was measured to make easier subsequent calculations and was evaporated almost dry under vacuum at 50°C. The residue was dissolved in a small amount of water, acidified to pH 2.0 by diluted HCl, and re-extracted 6 times by the peroxide-free ethyl ether (50 ml). Ether was evaporated, the residue was solved in 50% ethanol and passed through amberlite 401(CO_3^{2-}) and then eluted 1 N HCl in 50% ethanol. Ethanol eluate was evaporated to make easier chromatographic separation. Chromatography on the paper was conducted according to Kolombo and co-workers in descending order by n-butanol, saturated by 2% ammonia (it is recommended preliminarily to saturate paper with boric acid, to air dry and to recover to equilibrium with the water phase of solvent in the chamber). Two-dimensional chromatography: (1) 2% acetic acid; (2) chloroform-methanol-H_2O-formic acid (1000:100:200:8); water phase of the second solution is used for equilibrium during 24 hours before the second chromatography. The most successful is probably the method of descending single-dimensional chromatography in 2% acetic acid. Phenolic acids are found in ultraviolet or when spraying chromatogram with dinitrosulphoacid (DSA). Whitehead (1964) used CaO suspension in water (1:100) for the extraction of phenolic acids from different soils. He has succeeded in separating p-hydroxybenzoic, vanillic, p-coumaric and ferulic acids.

Kaminsky and Müller (1978) object to using alkaline ethanol extraction, because phenolic substances in this case are significantly changed and differ from phenols obtained with the use of Na-EDTA at pH 7.5 or in Soxlet apparatus with acetone.

Hence, the extraction of active substances was performed using different methods, but none of them was universal and compreh-ensive. Besides this, no method has tried to model the absorption ability of roots to know the amount of toxic substance taken up by plant. Therefore, a very important task is the development of such method of koline extraction, which would permit to obtain analytical and preparation amounts of kolines without soil destruction.

For soil investigations under comparable conditions, a special laboratory device was developed by O.D. Mironov. The main working element of the device is the column filled with the studied soil (column *b*), through which a solvent is filtered. The speed of filtration is regulated in a certain degree by efflux in the collector or by flow capacity of tube between the upper and bottom columns. The scheme of the device elements is presented in Fig. 1. Column ***b*** is densely filled with dry sieved soil (50-75 g) (the soil sample depends on the size of the column). The required extractant is poured into column ***a***. Its amount exceeds the soil mass 2-fold (i.e. the ratio between soil sample and extract is 1:2). Column *c* is a flask for the extract. All columns are hermetically connected. The extractant gets into the column with soil drop by drop and soaks through it. In case, when heavy clay soils are used, i.e. when the passing through is difficult for the extractant and can last even two days, the column ***c*** is connected to a vacuum pump to obtain uniform efflux. The speed of leaking is regulated by clamping screws coated on rubber connections. The collected water filtrates were cooled, if storage was necessary, but more frequently they were immediately used for analysis. If no further biotest is anticipated, it is advisable to add a drop of toluol or timol to exclude microbiological decomposition of substances.

This device can be used also for other solvents, except water. Only in the case of organic, slightly volatile, rubber joints were substituted for benzino-stable ones and instead of oil pump, in which oil is diluted and becomes worse, a water pump or other pump should be connected.

1.2.1. ***Comparison of extraction methods*** : We have conducted experiments on samples of typical soils taken from different zones of Ukraine and intensively degraded soil from forest and steppe hills and other places (Table 66). At first, the extraction by different solvents was compared.

Water and ethanol extracts appeared little active, precisely, their activity was in the range of stimulation. Methanol extracted approximately the similar amount of kolines from all the soils, whereas the rest of other solvents have major differences. Particularly many of active compounds were extracted with acidic extractants and the soil richer in humus keeps longer the high content of kolines in successive extraction. Therefore, the total activity of all extracted substances was highest in the low-moor peat–bog; a comparatively fertile soil from under beech and oak trees gave more substances, less – from under fir tree, and the least amount of substances was obtained from under the mountain pine tree.

Formic acid quite delicately hydrolyzes the substances of peat, soil from under fir tree and oak, whereas from the remaining soils - it extracts practically everything what they contain. One would think that extraction by acids is best than other solvents, but, there are many difficulties, which do not permit this method of extraction. (a) All acids cause unlimited hydrolysis of organic and mineral substances. If the elution has not been interrupted, only clean sand would remain in the column. To reflect the real absorption strength of roots, the acid extraction should be interrupted at some initial stage, but we do not know at which stage. (b) The applied acids are at first neutralized in the soil, the degree of neutralization being very different depending on buffer protection and on the presence of bases. This factor also prevents from working out common conditions for diverse oil extraction. In future, we should work with acid extraction, because it can model solvent strength of roots in full.

For extraction and chemical identification of separate kolines, the method should be based on solubility of their fraction. An example of this mode of extraction is given in Chapter 3.

Table 66. Content of physiologically active compounds extracted with different solvents (in relative coumarin units)

Sample origin	Water	Ethanol	Methanol	Sulphuric ether	1% acid			
					Formic	Hydrochloric	Acetic	Nitric
Oak, Zakarpacie	13.0	9.2	45.0	87.0	400.0	620.0	1100.0	22.9
Beech, Zakarpacie	6.4	6.4	55.2	11.8	900.0	665.0	760.0	4.9
Spruce, Ivano-Frankovskij region	3.0	8.4	55.4	17.0	275.0	760.0	600.0	120.0
Pine, Ivano-Frankovskij region	1.9	13.5	45.4	16.4	510.0	39.0	42.0	10.5
Peat bog, mouth of Souha river	12.5	5.6	60.1	15.0	510.0	1400.0	1400.0	1400.0

As it seems to us, the method of forcing out adsorbed substances from the soil is also very interesting. In this method a certain substance or a group of organic substances are applied to the studied soil and as a result of adsorption, these compounds displace the absorbed allelopathic substances, which are not extracted by usual methods. We used rutin (quercitin glycoside) as a substance, which was equally present in plant extracts, displaced absorbed allelopathic substances from soil. Air dry soil (50 g) was evenly mixed with 50 mg rutin and filled column *b* (see Fig. 27). Water and 50 % ethanol were let through column *a*. After extractant was passed through the column with soil and the introduced rutin or plant extract through the studied soil, the obtained extracts were evaporated. The residue was dissolved in 70% ethanol and chromotographed in n-butanol-acetic acid-water (40:12:28) system or n-butanol-acetic acid-water (3:2:95) on the paper. It appeared, that in comparison to the extraction with clean water and ethanol, which extracted nearly nothing from soil, and did not allow to find differences between the sick and control soils, the introduction of the «forcing out» substance (rutin) has made it possible to find and to identify certain kolines. Thus, the soil from under beech forest and from under Chinese aster has been found to contain quercetin. The same inhibitor was present in soil extracts from under beech and asters.

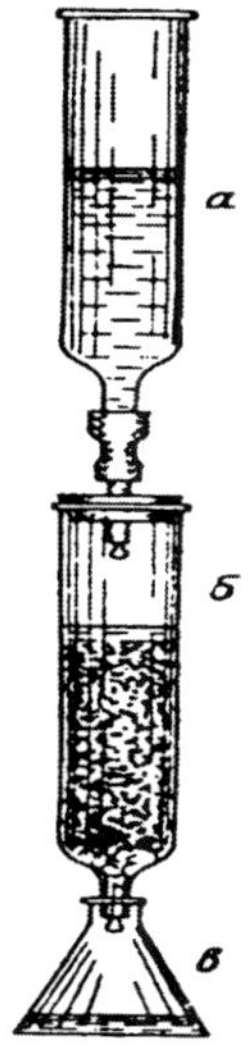

Fig. 27. Scheme of columns for extraction of active compounds from soil.

Table 67. Comparision of bioassay methods with other experiments

Experiment	Duration	Realization, relative units	Recurrence of results	Durability of examined compound	Grade of differentiation between control and experimental variants	Affinity to natural situation
Bioassay	30 min.- 48 h	300	Best	Good	Full	Least
Laboratory experiment (phytotron)	Some days – some weeks	10		Good in sterile condition	Good	
Vegetation experiment	Vegetation period	2		Weak, good with several application	Satisfactory	
Field experiment	One or more seasons	1	Least	Bad	Least	Best

Hence this method of koline extraction from soil is perspective. Especially important is the fact, that clear and convincing differences were found between the diverse soils, i.e. the specificity of certain sick soil was experimentally proved. Undoubtedly, it is extremely necessary to perform an entire series of further investigations to be convinced with this method. It is also desirable to test a number of other substances (probably also a mixture of compounds). We do not deny that our choice of the «forcing out» substance was not entirely successful, because rutin is phenol glycoside, and in this case more optimal will be aglycone to avoid decomposition of glycoside into its components. We have also tested other substances (*p*-benzaldehyde, gallic acid, coumarin), but the results were not satisfactory. Therefore, a detailed research is necessary in this direction.

Chemical interactions between the organisms constituting biocenosis (plants, animals and microorganisms) always involves a complex of very difficult problems. Some of them depend on the concentration and specificity in relation to acceptor plant, determine negative or positive effects of interactions. Others are precursors of these active substances or their derivatives, or simply play the role of synergists or antagonists of the main active compound, intensifying or weakening its effect and sometimes changing the sign of its activity. Even in cases, when the chemical nature of one or even many components is determined and their concentration is known, it is impossible to forecast the biological effect (Timiriazev 1949). The most acceptable for evaluation of the activity of substances extracted from the soils are biological methods (Table 67), though we cannot confine ourselves only to them, since chemical identification is also necessary.

2. BIOASSAYS

The biological methods based on determining the response of living organisms to this or other influence, are realized in field, vegetation and laboratory experiments and are called biological tests or bioassays. All mode of experiments are suitable also for allelopathic soil sickness, however, their effectiveness is different.

Bioassays are important in allelopathy, because it permits to exclude completely all factors, (which is not possible in field trial) are carried out quickly and avoids the destruction of investigated substances during the test. The bioassay provides maximum reproduction of the results and consequently, makes possible unification of investigations and a better comparison of the results of different tests conducted by different authors, even under different conditions.

However, the bioassay should be supported by other biological – vegetation and field methods. The chemical identification of excretions at all stages of purification should be controlled by the method of bioassays, otherwise, the compound "searched for" will be "lost" in one of intermediate stages. For the quickest determination of biological activity, such express-methods are suitable:

I. The speed of protoplasm movement in the epidermis cells of *Tradescantia*, in leaflets of *Elodea*, etc. (Roshtshina, 1975). Objects are placed in water and the investigated extracts are examined under a microscope with ocular micrometer.

II. The intensity of photosynthesis leaf discs in the atmosphere $^{14}CO_2$ (Mohova, 1962). Discs are placed in a chamber in light with a certain content of marked carbonic acid for 20 min, after which they are fixed, dried and radioactivity is estimated by the known methods. Discs are treated with the studied substances before placing them into the chamber or during photosynthesis – it is important to keep the other conditions equal in the control.

III. Exo-osmosis of ions after a change of water electroconductivity (Mohova, 1966). Seeds samples, seedlings, leaves and cut roots preliminarily treated with the examined substances are placed for 30 min in a 20-fold volume of distilled water and after that the electro-conductivity of initial water and samples are measured on any device with platinum electrodes.

IV. Exo-osmosis of optically active substances after a change of light transmission of water on the spectrophotometer (Mohova, 1966).

V. A change in the activity of enzymes, in particular proteases and respiratory enzymes.

VI. A change of interference or refraction of light waves. Carried out on corresponding instruments in air or water environment.

These tests require delicate handling with objects and some special equipment : (a microscope with an ocular micrometer, isotopes and devices with platinum electrodes for determination of electroconductivity etc.). The advantages of these tests are the rapidity and a suitable preparation of the results for mathematical calculations.

The allelopathic school, Moscow University uses the method of artificial free radical model (Yourin *et al.*, 1972). A water solution of 3,4-dihydroxyphenylalanine (DOPA) was used as a model in studying free radical changes in plant excretions. At oxidation of DOPA with atmospheric oxygen, it participates in free radical mechanism, whose final product is melanine.

However, these methods are less suitable for mass bioassays, which are most frequently conducted in allelopathic investigations, where biological activity of tens and even hundreds of samples is simultaneously compared. For these purposes it is better to use bioassays, which are called screening tests. These methods do not require a delicate handling and provide a mass statistical data available to mathematical calculation. Bioassays are recommended for seed germination and for initial growth of seedlings. Müller and his disciples (see Rice 1974) worked out a sand and sponge tests. In the sand test, the studied extract is introduced into a Petri dish containing sand of medium size and sow seeds of *Bromus rigidus*, *Festuca megalura* or of other plants. Several days after, the growth of seedlings is measured in comparison with the control on water.

In the sponge assay, cotton wool moistened with the tested extract was used as a substrate and seeds of experimental plants (*Bromus rigidus*, *Cucumis sativus* and others) were placed on this cotton wool between sheets of filter paper. When learning volatile excretions, the cotton is moistened with distilled water and near on both sides two small glasses with fragmented plant material are (1 g per each glass) kept (Müller *et al.*, 1964; Müller, 1966).

We used two bioassays described by us earlier (Grodzinsky, 1965; Grodzinsky, 1973) – for seed germination of radish and for radical growth of cress.

2.1. Bioassay of seed germination

It consists in counting the number of germinated seeds on the experiméntal solution, in comparison to the control germination on water, which constitutes 50%. We selected radish seeds of the cv. 'Red with a white end' as an experimental object, because out of all tested cultures they germinate the most rapidly and simultaneously. Usually during 24 hours and at the 26°C there is germination of 50% seeds and after some time – all remaining seeds. A high germination energy is desirable, since the more active are growing processes, the more sensitive is the plant organism to the influence of the environment and the shorter the time of a test, the lesser is the danger of microbial changes in the tested solutions. In addition to that, the shorter the time of a test, fewer plates, dishes and place space are required.

The tests have shown, that the same radish seeds, but with very low germination energy (old) did not respond to quite harmful substances. We use fresh, chosen seeds with high germination energy, from the last year of reproduction. To increase uniformity of seeds, they should be sieved through sieve (with mesh of 2.0-2.5 or 2.5-3.0 mm). Seed germination of radish was performed on filter paper in Petri dishes 9-10 cm in diameter. 200 seeds were sown per one dish. An optimal moisture is obtained by adding 5 ml of studied solution to the Petri dish. The amount of water 4-5 ml per dish has no significant influence on germination results.

Next the radish is sown, usually 600 seeds (3 Petri dishes per variant). Sowing should be done quickly; one hundred seeds should be counted before and placed in tube. Then the seeds are poured out and are evenly distributed on the paper. The test showed that seed counting in hundreds is not obligatory.

Radish seeds sown on Petri dishes germinate in darkness for 11-12 hrs at 26°C. The sowings are done at the end of the working day and the germination is determined in the morning of the next day.

A seed is considered to be germinated, when its radical broke the seed coat. Calculations should be made very quickly.

After that, an average seed germination of radish treatments is calculated in per cent to a corresponding germination on water, which is taken as 100%.

The higher is germination, it indicate less presence of germination inhibitor in the solution, at 100% - it is absent or its action is masked. Whereas at over 100% there are stimulators of germination. At 0%, the amount of kolines is extremely large. We should express the activity of the investigated solutions in relative units, namely in mg/l of coumarin - a known inhibitor accepted by as a standard. The activity of different coumarin concentrations is expressed by a typical one-peak curve with a full inhibition of germination at 1364 mg/l and stimulation at 5 mg/l. The testing of different coumarin concentrations has given quite coinciding curves. The value of a mean deviation from the standard curve between single measurements at 4-fold repeatability constituted 2.9%, which completely satisfies the requirements of precision in bioassays. The scale developed by us for recalculation of radish germination at K=50% into relative coumarin units is shown in Figure 28.

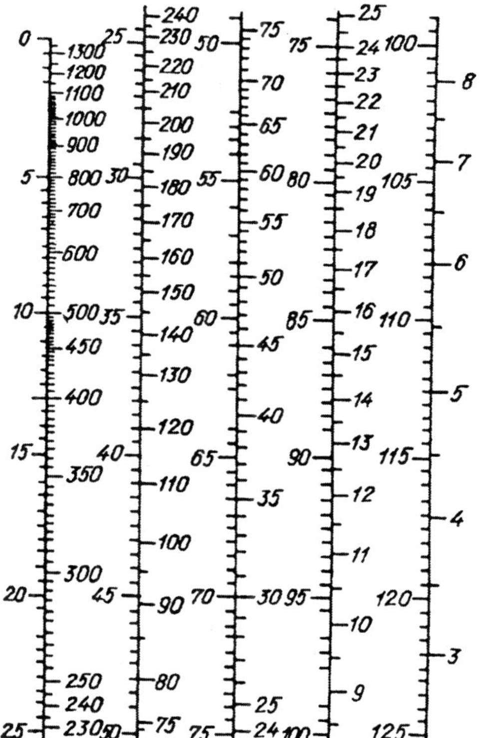

Fig. 28. Scale to transform data (in %) of radish seeds germination (on left) to relative coumarin units (on right).

For determination of volatile inhibitors, a sample of the corresponding material (about 3 g) on glass plates 4 x 6 cm² is placed into Petri dishes (do not touch a moistened filter paper on the bottom of the dish!). Seeds of radish are sown and estimation of germination is conducted as described above. Material sample (a source of volatile excretions) can be also placed on a boat made from aluminium foil, fastened to the wall of the dish, or in a small cup, or in crucible placed in the center. In this case, the dishes should be covered with lids of the same diameter and should be glued up with a sticking tape.

Bioassays with radish appeared to be highly reliable and stable, giving standard results even under deviating conditions. Owing to the subsequent recalculations radish germination at K = 50%, the influence of temperature fluctuation and of other conditions are levelled. Repeated testing of the same extracts at different time of the year on the seeds of radish of a different origin gave coinciding results. This permits us to recommend the above bioassays, not only for physiological, but also for experimental geobotanical investigations, under conditions of expeditions and weakly equipped laboratories, preserves and stationeries. Bioassays on seeds can be applied also by agricultural workers for determination of kolines in the fruits and seeds, post-harvest residues, in local fertilizers when studying the problems of soil sickness, etc. These tests appeared to be more reliable than the known bioassays on coleoptiles, root parts and they were also much more productive.

2.2. Bioassay with seedling growth

Sorted out seeds of cress are sown on the filter paper moistened with distilled water and are placed in a dark thermostat at 27°C. After 24 hrs, when the seeds germinate, 3-5 mm radicals are taken to place them into Petri dishes (20 radicals per dish) with the tested extracts. After the next 24 hrs, the root length is measured and expressed in % to the root length of control in water. It is more convenient to measure roots by photographical method (Grodzinsky and Grodzinsky, 1973).

It is recommended to use the test specially developed by Poptzov (1959) for determination of soil toxicity. The essence of this method is as follows. Seed germination is done in the investigated soil and clean sand (washed and tempered) in Petri dishes, pots or wooden boxes. Seeds of white and black mustards (mustard or other

Cruciferae is highly sensitive to herbicides and other inhibitory substances) and wheat (distinguished by the highest stability to toxic products) are used for the tests. The germinating ability is determined after 2-3 days. The degree of radical development in soil and sand and deviation from normal development (deformity, rotting, turning black etc.) are determined visually. Of course, a test culture should be selected taking into account a supposed use of soil.

With the aim to intensify phytopathogenic effects of soil, the test samples are placed for some time into growth chambers at 1–10°C. Cucumber plants at this temperature in sterile soil did not suffer, but in the presence of fusariose or other infection they died even if it returned to warm conditions.

To study phytopathogenicity of various soils to sugar beet, a special test has been developed in East Germany. It can be also used for any plants. Plants are grown for 20 days in 100 ml plastic pots joined on a common perforated board consisting of 51 pots. This board is fixed into the same one, and with the use of wooden washers is supported over it at the height of 2–3 cm. The upper pots have the holes in the bottoms, which are closed by drainage (glass wool etc.). Each pot contains 50 ml of the studied soil, into which one seedling of beet or other plant is planted. Water, or a solution of nutrients, insecticides, fungicides or herbicides, is poured out into lower pots up to a definite level (when learning soil moisture, the level of water varies within necessary limits). Plants are watered through a regular addition of glasses with water to a necessary level. For this purpose the board is lifted first for 3-4 cm to let the water to flow down from the upper pots. After that, it is taken away.

The use of perforated board is comfortable, because group of 51 plants is concentrated and easily transported, i.e. a small test with 6 variants at 8-fold repeatability can be placed on one board. When there is no perforated board, it is better to place pots in groups on the common board on saucers with water and solutions. It is probably suitable also for screening fungicidal preparations, if pots are filled with deliberately infected soil, introducing tested substances into each pot. The last two bioassays belong to laboratory tests, since their duration exceeds several days.

Neubauer-Schnaider (Grodzinsky and Grodzinsky, 1973), method is suitable for the determination of the presence of mineral elements in soil. Seeds of rye are sown into glass or plastic crystallizators (11.0–11.5 cm in diameter and 7 cm high), filled with a mixture of studied soil and sand. Short tubes for watering are placed into the centre of crystallizers. The duration of the test is 14 days. Some authors used it for screening the effect of derivatives of thimol hydroquinone, products of straw decomposition.

Besides that, there are laboratory tests for volalites, which give a clear visual presentation about the influence of active substances and other factors (Molisch, 1937), but which are more difficult in mathematical presentation:

(a) Epinasty of vetch and pea plants in pots, (Molisch, 1937); recommended this test to display fire-damp in mines.

(b) Acceleration of leaf falling from cut branches of pea tree put in water under a glass bowl, where the investigated volatile substances are introduced, leaves of the control fall approximately a week later.

(c) Leaf turning yellow in darkness (response to cytokinins).

H.G. Cholodny used very simple equipment for studying volatile phytogenous substances of soil: glass vessels were turned over and put on the side obliquely, in which slightly moist soil or swollen seeds were placed. In a similar manner, Persidsky and Wilde (1954) studied the volatile substances of composts and leaf litter. Microscopic slides were strengthened across a glass vessel, on which cut roots of alfalfa, sunflower or maize were distributed on a wet filter paper, or a thin layer of soil powder was put on a dry slide. The vessel was covered by a glass plate put from the side. It is recommended to air the vessels twice a day by opening them for some seconds. After 2 days, roots in the control vessels begin to die off; bacterial colonies do not develop on the soil powder. In the presence of phytogenic organic substances, the life of cut down roots is extended for 3-4 days; even dry matter increases in them by 25%. Abundant colonies of bacteria feeding the organic volatile substances released from the soil to the air develop on the microscopic slides with soil powder. It seems to us that these bioassay conducted long ago deserve to be continued and developed.

When investigating allelopathic soil sickness in apple and other fruit plants, P.A. Morozov, I.I. Popivschi and V.I. Gorinova in our laboratory used a test on shoots of apple and other plants kept in tubes on a nutrient solution supplemented with the tested extracts, or grown in sand culture in small porcelain vessels, where hard residues (leaves, roots) or extracts from them were added. Apple shoots appeared to be very sensitive to allelopathically active substances and responded by stopping stem and root growth, leaf withering and turning brown.

Bioassays are generally based on very different physiological responses, two tendencies being of predominant importance. One of them pursued mastering of the most clear specific response of test-object to only one active substance, independently on other compounds present in the solution (substrate). The first researchers of growth substances have considered for 25-30 years that there is only one substance – indolilo-3-acetic acetic (IAA) having an influence in bioassays with plant extracts. However, later on, it appeared that many other substances, different from IAA, have a similar influence on biological objects, for instance, sugar substances and organic acids. Many researchers were disappointed, which caused distrustfulness to bioassays. To avoid unspecificity, investigators started seeking for refined tests using only one physiological response. This has resulted in some success, though there are only a very few such responses in the arsenal of bioassays (bulky test for oat coleoptiles bend, test for preservation of chlorophyll protein complex of leaves in darkness under the influence of cytokinine test for increase of amylase activity under the influence of gibberellins).

Another possibility of creation of specific bioassays consists in using materials, which genetically are unable to synthesize some substances. Thus, there are fungal races which cannot synthesize vitamins B, or neurine. No other substances will force the fungus to grow, until will receive these vitamins or their nearest derivatives. Some isolated tissues, which cannot synthesize certain growth substances, display quite a high specificity.

3. MICROBIOLOGICAL TESTS

The use of microbiological tests is of interest for determination of allelopathic soil sickness. Because microbiological populations at soil sickness undergo quite strong changes, in particular their species composition is impoverished; it may be supposed that there are races and species of microorganisms that do not induce soil sickness.

In toxic soil, not only cultured plants do not want to grow, but also very often many useful species of microorganisms die including nitrobacteria, cellulolytic, nodular and oligonitrophilous microorganisms. Microbial balance is disturbed, the relations among microorganism groups are shifted, the number of bacillary forms of bacteria, actinomycetes, fungi, among which a large number of toxin-producers and antagonists to useful species of microorganisms were found. Phytopathogenic microflora, which suppresses development of saprophytes also frequently develops. Of course, depending on the type of soil, sickness and many other conditions, intolerant species can be different and it is difficult to find a microbiological test universal for all soils.

A number of microbial test objects are described in the literature. They can be used when elucidating the degree of soil toxicity. Muromcev and co-workers applied microscopic water-plant *Chlorella vulgaris* as a test organism, the lyses zone of which around the colonies of bacteria and fungi indicates the production of toxic substances by them.

Good indicators of soil toxicity are organisms decomposing the cellulose (Jukova, 1964). Berestecki (1969), when explaining the causes of toxicity under orchard trees and in monoculture of different plants equally with plant tests used also microbe - gram-positive and gram-negative microorganisms.

According to the data of Rice (1974) and numerous sources mentioned in his book, the allelopathic soil sickness, which occurs under diverse herbal and woody plants, one of the most common responses is the suppression of nitrogen fixation of free-living and symbiotic organisms and also conversions of ammonia and amide compounds into nitrates i.e. nitrification. This is related with the suppression of corresponding microorganisms.

Our studies in sick soil (under monocultures of flower, fruit and agricultural plants) have shown that nitrobacteria, were absent. Taking into account this fact and the paper by Roizin (1969), we decided to use nitrobacter (*Azotobacter chroococcum)* as a test-object. The method of the investigation is as follows: soil moistened to 60%, in which silk threads preliminarily sterilized and inoculated with *Az. chroococcum* were sustained from 5 to 25 days, was put into Petri dishes. After that, the threads were taken away and put into Petri dishes with the Ashbi medium. The intensity of nitrobacter development was measured according to a 5-degree scale, which indicated the degree of soil toxicity (the higher the toxicity, the lower the degree). The selection of test objects to determine the effect of sick soil is important. A high degree of toxicity was well displayed in the plant bioassay, especially in test of cress radicals.

Nitrobacter test is a more universal bioassay - its sensitivity is displayed not only in very sick soils, but also in soils with a medium and low toxicity. This test object is sensitive not only to soil sickness connected with monoculture of field and orchard cultures, but is also sensitive to high doses of mineral fertilizers.

When using different test-objects for estimation of soil toxicity, it should be mentioned that soil toxicity in the dynamics is a variable value, which is connected with the rhythms of season and diurnal activity of microflora. For this reason, it is necessary to calculate an average index of soil toxicity.

Chapter 9

CONCLUSIONS

To take care of soil fertility and its improvement is the most important tasks of mankind. Here are three main problems of the modern scientific technical revolution: (a) A search for the sources of energy – inexhaustible, if possible, harmless and suitable in exploitation. (b) Solution of the food problem of the steadily increasing population of the Earth. (c) Preservation of ecological balance in the biosphere of the Earth.

These three cardinal problems should be solved for successful existence and development of the mankind civilization. All these problems are related to soil fertility. Even the energy problem, despite enormous investments in the construction of nuclear power stations, is solved mainly by photosynthetic plant products i.e. food grain etc.

The role of soil fertility to solve the food problem does not require any commentary. The introduction of semi-dwarf wheats in Mexico, Pakistan, India and in other countries of Central America, the Near and Middle-East was called "green revolution", as it sharply reduced the food deficit.

To maintain ecological balance in the biosphere, fertile, actively photosynthesizing fields are of great importance, since they absorb and process carbonic acid, nitrogen oxides and other various waste materials of industry, transport and human activity, making the environment healthier. Hence, the soil productivity is major condition of mankind existence. Therefore, soil sickness as a process decreasing the soil productivity and crop yields deserves great attention.

In the previous chapters, we have considered some questions of allelopathic soil sickness, i.e. accumulation in soil of organic substances having a negative influence on plant growth. If we compare the ancient and modern systems of agriculture, it is evident that in modern agriculture we are reducing the content of organic substance (humus) in soil. For this purpose trees and bushes, and peat was burnt out. In steppe zone, farmers did not know the fertilizers. In Ancient Rome, only on well ploughed areas, organic fertilizers – humus, compost, green manure were initially used. Now a days, crops cannot be grown without inorganic fertilizers, which are most effective and meets the requirements of plants. This creates an impression that an optimal yield requires a certain content of organic substance in soil. When it is too much, a farmers strive to reduce it by mineralization, whereas when it is a little or not enough, farmers specially apply organic fertilizes into a soil. All this has been found purely empirically, on the basis of experience of thousands of millions of farmers, but not by strict scientific investigations.

Until now, the role of soil organic substance has not been probably elucidated to the end. It has been scientifically proved, that plants need sun light, water, carbon dioxide and a certain complex of macro- and microelements to produce yield and there is no necessity to supply them organic products. Because of that soil organic substance (humus) is studied only as a colloid joining soil particles together and providing this or that soil structure. The experience of agriculture of many centuries indicates that it is impossible to obtain optimal yield on the exhausted soil without applying fertilizers.

The plants can live in completely mineralized conditions, in practice they always encounter the presence of larger or smaller amounts of mobile, available organic substances in substrate. Increased concentrations of these substances leads to soil toxicity and inhibit growth.

Therefore, the main task of future agriculture is to learn to regulate the level of free organic substance (and, of course, its qualitative composition as well) to avoid soil sickness and to obtain high yields on arable lands, to solve the three main problems of mankind viz., energy, food and ecology.

Another very interesting, but not scientifically explained phenomenon closely connected with the problem of soil sickness, is stimulation of crops yield after soil sterilization. Advancing different hypotheses of the reasons of soil sickness, people again and again returned to the idea about the necessity of soil sterilization before sowing to destroy nematodes, arthropod pests, bacteria and fungi, sprouts of weed plants, etc. A poet of Ancient Rome - Virgil describing in verse a rich experience of agriculture in Rome of those times wrote: "Burning of exhausted soil is useful to it "...flame will burn any defect..." (translation). Later conducted experiments on soil sterilization by warming it thoroughly up to 200-250°C by autoclaving, steaming, irradiation with gamma or X-rays, application of toxic substances, such as chloropicrin, hydrocyanic acids and other toxins, which do not remain in soil as a result of instability or volatility, almost always led to a better growth and to an increase in higher plant yields. High prices and technical complication of soil sterilization and some other problems difficult to overcome, did not make it popular. Practically it is impossible to sterilize soil without destruction. The soil will always have spores of bacteria and fungi, capsules of water plants and stable embryos of other organisms. After the end of sterilization, the soil is quite rapidly inhabited. In this case, we can speak only about a partial or provisional sterilization.

Strictly speaking, an every year tillage with a layer turn over is also a sterilizing treatment, because the microflora strongly developed in the upper layer dies, found themselves on the furrow-slice bottom, and deep soil brought upward contains much less microorganisms and achieves their previous activity only after one or one and a half month.

REFERENCES

Aamisepp A., Osvald H. 1962. Influence of higher plants upon each other – allelopathy: Some new results of research into allelopathy. *Nova Akta Regiae Soc. Sci. Upsal.* **2**: 1-19.

Abdul-Wahab A.S., Rice E.L. 1967. Plant inhibition by Johnson grass and its possible significance in old-field succession. *Bull. Torrey Bot. Club* **94** (6): 486-497.

Aldag R. 1966. Soluble Amino Acids of Soil and Their Effect on Higher Plants. Diss. Göttingen. (In German).

Alieva E.I. 1975. Use of straw as nutrition. *Agrochimia* **2**: 96-99. (In Russian).

Allen M.V., Lensen H.J. 1951. *Pratylenchus vulnus,* new species, nematode, pratylenchinae and parasite of trees and vines in California. *Proc. Helminthol. Soc. Wash.* **18**: 39-45.

Al-Mousawi A.H., Al-Naib F.A.G. 1976. Volatile growth inhibitors produced by *Eucalyptus microtheca. Bull. Biol. Res. Cent.* **7** (1): 17-23.

Altimirskaya P. 1976. Microbiological investigation of soil in wheat monoculture. *Potshvoznanye i Agrochemya* **11** (1): 90-95. (In Russian).

Andrieyev K. 1972. Useful advice. *Tsvietovotstvo* **5**: 29. (In Russian).

Apelbaum A., Burg S.P. 1971. Altered cell microfibrillar orientation in ethylene-treated *Pisum sativum* stems. *Plant Physiol.* **48** (5): 648-652.

Aristovskaya T.W. 1975. General criteria of estimation of soil microflora role in ecosystem functioning. In: *Ecology of Microorganisms. Proc. 5th Soviet Conf. Microbiological Soc.,* Erevan p. 3-4. (In Russian).

Baeumer K., Lochov. I.V. 1961. Study of real soil sickness by clover. *Z. Acker und Pflanzenbau* **113** (4): 395-412. (In German).

Baines R.C., Clarke O.F. 1952. Citrus-root nematode effect on young lemon and orange trees studied in inoculation tests under controlled conditions. *Calif. Agr.* **6**: 8-13.

Balabanov M.S. 1908. Improvement of old apple orchards. *Progressivnoye sadovodstvo i ogorodnitshestvo* **38**: 481-482. (In Russian).

Balicka N., Trzebinski M. 1956. Activity of enzymes and occurrence of vitamin B_2 in soil. *Acta Microbiol. Polon.* **5**: 377-384.

Ballantyne D.J. 1962. A growth inhibitor from the roots of *Chrysanthemum morifolium* Ramat and its influence upon the giberellic acid response. *Can. J. Bot.* **40** (9): 1229-1235.

Baxter P. 1961. Propagation of apple in sick soil. *Eewerds-Obstbau* **1** (3): 1-2. (In German).

Bear I.J., Thomas R.G. 1964. Nature of argillaceous odour. *Nature* **201**, (4923): 993-995.

Bear I.J., Thomas R.G. 1965. Petrichor and plant growth. *Nature* **207**, (5004): 1415-1416.

Beck Th. 1975. Effect of perennial monocultures on exhausting soil in comparison to crop rotation systems. *Landwirt. Forsch.* **31**: 268-276. (In German).

Bekker Z.E. Lebed E.S. 1966. Some data on dynamics of antagonistic fungi in soil under monoculture of cotton plants. *Nautshnoye Doklady Wyshey Shkoly*. Ser. B, **3**: 184-189.

Beltukova K.I., Lapyonov D.K., W.M., Maculevitsh B.P., Salunska N.I. 1959. *Diseases of Perennial Legumes and Their Diagnoses.* Diertshsilgosvidav URSR 179 pp. (In Ukrainian).

Berestecki O.A. 1967. To question about reason of toxicity some soils of Polesie. In: *Nitrogen Problem and Yields on Polesie*, Kiev p. 31-33. (In Russian).

Berestecki O.A. 1969. Changes of soil microflora by fruit plants in connection with toxicity of orchard soils. *Proc. I[st] International Conference on Agrophytocenology.* Kazan University p. 161-167. (In Russian).

Berestecki O.A. 1970. Role of products of root residue decomposition in toxicity of orchard soils. *Physiological and Biochemical Basis of Plant Interactions in Phytocenoses* **1**: 113-118. (In Russian).

Berestecki O.A. 1970. Soil toxicity from under apple orchars. In: *Soil Conditions, Nutrition and Fertility of Fruit Cultures.* Kiev, Urotshay p. 477-484. (In Russian).

Berestecki O.A. 1971. Production of phytotoxic compounds by soil micro-organisms and their role in allelopathy. *Physiological and Biochemical Basis of Plant Interactions in Phytocenoses* **2**: 125-132. (In Russian).

Berestecki O.A. 1971. Soil toxicity under perennial fruit trees cultures. *Potshvovyedyenyie* 7: 56-59. (In Russian).

Berestecki O.A. 1972. Effect of toxins from *Penicillium claviforme* Bainier and *Penicillium brevi-compactum* Dierckx. *Microbiol. J.* **34**(3): 377-381. (In Ukrainian).

Berestecki O.A. 1972. Production of phytotoxic compounds by soil micro-organisms on root residues of fruit plants. *Physiological and Biochemical Basis of Plant Interactions in Phytocenoses* **3**: 121-124. (In Russian).

Berestecki O.A. 1974. Phytotoxic properties microbial fungi obtained from rhizosphere of old apple trees. *Mikologya i Phytopatologya* **8** (1): 69-72. (In Russian).

Berestecki O.A. 1974. Toxicity of orchard soils in dependence on ecological conditions and biological properties of fruit plants. *Physiological and Biochemical Basis of Plant Interactions in Phytocenoses* **5**: 47-50. (In Russian).

Berestecki O.A. 1975. Micro-organisms as factor of soil toxicity under perennial fruit cultures. *Potshvovyedyenyie* **4**: 92-97. (In Russian).

Berestecki O.A. 1976. Role of micro-organisms in allelopathy. Proceedings of 4th Soviet Union Symposium on Physiological and Biochemical Basis of Interactions of Plants in Phytocenoses. Naukova Dumka, Kiev pp. 4-5. (In Russian).

Berestecki O.A., Kalmykova N.A. 1974. Micro-organisms – factor of turf-podzolic soil toxicity in monoculture of winter wheat and maize. *Microbiol. J.* **36** (6): 726-731. (In Russian).

Berestecki O.A., Moroz P.A. 1968. Effects of some factors on phlorizin content in apple roots and leaves in connection to soil sickness. *Agrochimia* **6**: 66-69. (In Russian).

Berestecki O.A., Patyka W.F. 1976. *Chemical Nature of Phytotoxic Compounds of Some Soil Micro-organisms.* Naukova Dumka, Kiev pp. 13-14. (In Russian).

Berestecki O.A., Sinitzki W.N. 1973. Phytotoxic activity of various strains of *Penicillium urticae* Bainer. *Microbiol. J.* **35** (3): 349-353. (In Russian).

Berestecki O.A., Tortshetski W.I., Motshalov J.N. 1976. Curiosities of microflora of turf-podzol soil in crop plant monoculture and crop rotation systems. *Mikrobiologia* **15** (4): 710-717. (In Russian).

Bilay W.I. 1961. *Microscopic Fungi – Producers of Antibiotics.* Edition of AN USSR, Kiev, 181 pp. (In Russian).

Bisovietsky T. J., Kotsurenko M. H., Fiedorenko L.W. 1966. Roots and post-harvest residues of beet crop rotation system and their chemical composition. *Agrochimia* **9**: 83-87. (In Russian).

Blagonravova L.N., Shtsherbanovsky L.P. 1975. Effect of 1,4-naftoquinones and saponines on polyphenol oxidase activity. *Physiological and Biochemical Basis of Plant Interactions in Phytocenoses* **6**: 37-39. (In Russian).

Blake M.A. 1947. The growing of peaches after peaches was an old world problem in 1768. *New Jersey Hort. News* **28** (1): 1910.

Bode H.R. 1958. Allelopathic symptoms occurring in some plants of *Juglandaceae. Planta* **3**: 440-480. (In German).

Bogdan G.P. 1971a. Anatomical studies of damages of plant root system by allelopathically active substances. *Ukr. Bot. J.* **28** (6): 703-707. (In Ukrainian).

Bogdan G.P. 1971b. Plant response to influence of allelopathically active compounds. In: *Physiologically Active Substances with Biogenic Origin.* Edition of Moscow University, p. 94-96. (In Russian).

Bogdan G.P. 1971c. Chlorogenic acid and their importance in defense responses of plants. *Physiological and Biochemical Basis of Plant Interactions in Phytocenoses* **2**: 119-124. (In Russian).

Bogdan G.P. 1972a. Histochemical determination of phenolic compounds in acceptor-plants. Physiological and Biochemical Basis of Plant Interactions in Phytocenoses. *Proceedings 3rd Soviet Union Symposium,* Naukova Dumka, Kiev, p. 43-44. (In Russian).

Bogdan G.P. 1972 b. On chemical nature of substances caused corking of xylem vessels influenced by active allelopathic compounds. *Physiological and Biochemical Basis of Plant Interactions in Phytocenoses* **3**: 62-63. (In Russian).

Bogdan G.P. 1974 a. Importance of peroxidase activity in defense response of plants induced by allelopathic factor. *Physiological and Biochemical Basis of Plant Interactions in Phytocenoses* **5**: 19-21. (In Russian).

Bogdan G.P. 1974 b. Effect of allelopatically active compounds on pH of tissue environment of acceptor-plants. *Physiological and Biochemical Basis of Plant Interactions in Phytocenoses* **5**: 21-23. (In Russian).

Bogdan G.P. 1976. Effect of allelopathically active compounds on activity of non-specific dehydrogenases in acceptor-plants. In: *Problems of Allelopathy.* Naukova Dumka, Kiev p. 14-15. (In Russian).

Bogdan G.P. 1977. Determination of enzymes participating in defense responses of couch-grass with allelopathic activity. In: *Relationships of Plants and Micro-organisms in Phytocenoses*. Naukova Dumka, p. 29-36. (In Russian).

Bogdan G.P., Grodzinsky A.M. 1973. Participation of sulphhydryl groups in histogenic plant responses to allelopathical damages. *Ukr. Bot. J.* **30** (6): 77-778. (In Ukrainian).

Bolotov A.G. 1952. *Selected Works*. Moscow Society of Nature Defenders 523 pp.

Bonner J. 1946a. Relation of toxic substances to growth of guayule in soil. *Bot. Gaz.* **107** (3): 343-351.

Bonner J. 1946 b. Further investigation of toxic substances which arise from guayule plants: relation of toxic substances to the growth of guayule in soil. *Bot. Gaz.* **107** (3): 343-351.

Bonner J. 1950. The role of toxic substances in the interactions of higher plants. *Bot. Rev.* **16** (1): 51-65.

Bőrner H. 1958. Determination of phloridzin realizing in soil. *Naturwissenschaften* **45** (6): 138-139.

Bőrner H. 1959. The apple replant problem. I. The excretion of phlorizin from apple root residues. *Contr. Boyce Thompson Inst.* **20**: 39-56. (In German).

Bőrner H. 1960. Experimental trials on soil sickness of apple (*Pyrus malus*). *Beiträge Biol. Pflanzen* **36** (1): 97-137. (In German).

Bőrner H. 1960. Liberation of organic substances from higher plants and their role in the soil sickness problem. *Bot. Rev.* **26**: 393-424.

Bőrner H. 1961. Interrelations of higher plants. *Naturwiss. Rdsch.* **14** (7): 265-269. (In German).

Bőrner H., Martin P., Clauss H., Rademacher B. 1959. Studies on soil sickness problem on example of flax and rye. *Z. Pflanzenkrankh. Pflanzenschutz* **66** (11/12): 681-703. (In German).

Bőrner H. 1963. Study of phytotoxic and antimicrobial substances formation in soil and their possible importance in soil sickness of apple (*Pyrus malus* L.). Formation of patuline and phenolic compounds by *Penicillium expansum* on apple roots and fallen leaves. *Phytopathology* **48** (4): 370-396. (In German).

Bremner J.M., Harada T. 1959. Release of ammonium and organic matter from soil by hydrofluoric acid and effect of hydrofluoric acid treatment on extraction of soil organic matter by neutral and alkaline reagents. *J. Agric. Sci.* **52** (2): 91-94.

Bronsart H. 1949. Current state of our knowledge on soil sickness. *Z. Pflanzenernarh. Bodenk.* **45** (2): 166-193. (In German).

Brown K.F. 1964. The assessment of a plant parasite nematode problem and a recent development in methods for chemical control. *Proc. 8th Int. Congr. Soil Sci.*, Bucharest **3**: 58-60.

Bublitz W. 1959. On organic inhibito... and bioassays. *Naturwissenschaften* **40**: 99-102.

Budagovsky W.I. 1976. *Culture of Weakened Fruit Trees.* Kolos 300 pp. (In Russian).

Bukolova T.P. 1969. Koline effect of sow thistle *Sonchus arvensis* L. and pigweed *Chenopodium album* L. on anatomical structure some crop plants. In: *Soviet Union Works on Study of Plant Relationships in Phytocenoses.* Nauka i Technika, Minsk p. 160-162. (In Russian).

Bukolova T.P. 1971. Antimitotic effect of water extracts from above ground plant parts and roots of some weeds. In: *Physiological Active Compounds with Biogenic Origin.* Edition of Moscow University p. 98-100. (In Russian).

Burangulova M.N. Rahimov E.M., Haziyev F.H., Yanbuhtina R.N., Sahabutdinova A.Z. 1974. Effect of crop rotation and monoculture systems on microflora and some biochemical processes in exhausted chernozems. In: *Soils of Bashkiria,* Ufa p. 204-216. (In Russian).

Byeltyukova K.I., Lapyonov D.K., Lopatin W.M. 1959. *Diseases of Perrenial Legumes and Their Estimation.* Deptshsilgospvidav USSR, Kiev. (In Ukrainian).

Candolle de A. P. 1813. *Elementary Theory of Botany.* Paris. (In French).

Cannon J.R., Corbett N.H., Haydock K.P., Tracey J.G., Webb L.J. 1962. An investigation of the effect of the dehydroangustione present in the leaf litter of *Bachkousia angustifolia* on the germination of *Araucaria cunnighamii* – an experimental approach to a problem in rain-forest ecology. *Austral. J. Bot.* **10** (2): 119-128.

Chang C.-F., Suzuki A., Kumal S., Tamura S. 1969. Chemical studies of " clover sickness". II. Biological functions of isoflavonoides and their related compounds. *Agr. Biol. Chem.* **33** (3): 396-408.

Chou C.-N., Muller C.H. 1972. Allelopathic mechanism of *Arcostaphylos glandulosa* var. *zacaensis. Amer. Midl. Naturalist* **38** (2): 324-347.

Choudri M.B., Stevenson F.J. 1957. Chemical and physicochemical properties of soil humic colloids. 3. Extraction of organic matter from soils. *Proc. Soil Sci. Soc. Amer.* **21** (12): 508-513.

Collina F., Zucconi F. 1967. Input in studies on effects of post-harvest residues of different crops on apple tree growth. *Riv. Ortoflorofrutticolt. Ital.* **51** (5): 386-406. (In Italian).

Conrad J.P. 1942. The occurrence and origin of urease-like activities in soils. *Soil Sci.* **54** (5): 367-380.

Coupin H. 1911. *Comparision of Plant Secretion Toxicity.* Acad. Sci., Paris, 152 pp.

D' Herde J., MacSenner J., Brande J. 1961. Soil sickness in begonia culture. *Meded. Landbouwhogesch. Opzoenhingsstas stoat. Gent,* **26** (3): 71-74. In Dutch.

D'Herde J., Brande J. 1962. Soil sickness in nurseries and rose field in Belgie. *Meded. Landbouwhogesch. Opzoenhingsstas stoat. Gent,* **27** (3): 768-775. In Dutch.

Danilevsky A.F., Yeshchenko W.E. 1972. Accumulation of plant residues of crop cultures and binding of nutrient elements in their. *Agrochimia* **8**: 65. (In Russian).

Daubeny C.G.B. 1845. On the rotation of crops and on the quantity of inorganic matters abstracted from the soil by various plants under different circumstances. *Phil. Trans.* **1845**: 179-252.

Dietz R. 1959. Effect of aminoacids on plant growth. *Bodenkultur* **10** (2): 144-156.

Dospiehov B.A. 1963. Half-century study on monoculture of flax. *Izv. Timiriazev. S.-H. Akad.* **53** (4): 33-41. (In Russian).

Dospiehov B.A., Vasilieva D.B., Kurienkova I.K. 1971. Soil fertility and yielding of fibre flax in long-lasting monoculture. *Izv. Timiriazev. S.-H. Akad.* **61** (3): 31-39. (In Russian).

Druckrey H., Schildbach A. 1963. Quantity determination and importance of benzopyrens in cancerigenic effect of cigarette smoke. *Z. Krebsforsch.* **65** (4): 465-472. (In German).

Dubach P., Mehta N.C. 1963. The chemistry of soil humic substances. *Soil and Fert.* **26** (3): 293-300. (In German).

Dubach P., Mehta N.C., Deuel H. 1963. 1963. Economical extraction of humic substances and isolation fulvic acids. Fractions from different kinds of soil. *Z. Pflanzenernahr. Bodenk.* **102** (1): 1-7.

Dublslav G. 1966. Introduction of Crop Rotation Systems with Consideration of Local Conditions. Kolos, Moscow 18 pp. (In Russian).

Dzubenko N.N. 1966. Accumulation of water-soluble kolines under pulses and cereal crops. In: *Physiological and Biochemical Basis of Plant Relationships in Phytocenosis.* Nauka, Moscow p. 125-130. (In Russian).

Dzubenko N.N., Boyko P.I. 1972. Some results of study of forecrop role on koline dynamics in crop rotation system. *Physiological and Biochemical Basis of Plant Interactions in Phytocenoses* 3: 55-58. (In Russian).

Dzubenko N.N., Krupa L.I., Boyko P.I. 1977. Dynamics of accumulation of pests in monoculture and crop rotation soils. In: *Relationships of Plant and Microorganisms in Phytocenoses*. Naukova Dumka, Kiev p. 70-78. (In Russian).

Dzubenko N.N., Sigaryeva D.D., Boyko P.I. 1973. Effect of forecrops on koline and phytonematode dynamics in crop rotation system. *Physiological and Biochemical Basis of Plant Interactions in Phytocenoses* 4: 38-44. (In Russian).

Elisashvili T.A. 1976. Effect of phytotoxic metabolites of *Myrothecium* sp.fungi on sprouts, leaves and cells of plants. In: *Problems of Allelopathy*. Naukova Dumka, Kiev p. 42-43. (In Russian).

Ellwardt P.-Ch. 1977. Variation in content of polycyclic aromatic hydrocarbons in soil and plants by using municipal waste composts in agriculture. In: *Soil organic matter studies*. Vienna: IAEA, vol. 11, p. 219-298.

Fastabend H. 1955. Soil sickness – deficiency of micro-elements? *Gartenwelt* 55 (9): 137-139. (In German).

Fedin M.A., Biter A.F. 1960. Monoculture of maize. *Kukuruza* 7: 19-20. (In Russian).

Flaig W., Beutelspacher H., Reitz E. 1975. Chemical composition and physical properties of humic substances. In: *Soil Components*. New York: Springer, p. 1-211.

Forster R.C., Rovira A.D. 1976. Ultrastructure of wheat rhizosphere. *New Phytol.* 76 (2): 343-352.

Friedrich G. 1970. *Fruit Farming*. Neumann, Leipzig, 843 pp. (In German).

Galstyan A.Sh. 1961. Soil respiration as one of symptoms its biological activity. *Dokl. AN ApmSSR* 33 (4): 33-34. (In Russian).

Galstyan A.Sh. 1962. Method of determination of dehydrogenase in soil. *Dokl. AN ApmSSR* 35 (4): 181-184. (In Russian).

Galstyan A.Sh. 1963. To learning biological activity of soil. *Dokl. AN ApmSSR* 37 (2): 89-93. (In Russian).

Galstyan A.Sh. 1965. Method of determination of hydrolytic enzymes. *Potshvoviedienije* 2: 68-74. (In Russian).

Galstyan A.Sh. 1968. Some problems of soil enzymatology. *Proc. Symposium on Soil Enzymes*, Minsk p. 24-38. (In Russian).

Gamzikova O.I. 1968. Effect of crop plants on enzymatic activity of soil. *Proc. Symposium on Soil Enzymes*, Minsk p. 327-333. (In Russian).

Gascho G.F., Stevenson F.J. 1968. An improved method for extracting organic matter from soil. *Soil Sci. Soc. Amer. Proc.* **32** (1): 117-118.

Gerkyal A.M. 1974. Accumulation of organic compounds and mineral elements in soil from post-harvest and root residues of fallowing cultures. *Agrochimia* **2**: 15-20. (In Russian).

Gil L.S., Zykova T.A., Kaplan C.M. 1968. *Gladioluses. Dahlias. Cannas.* Urozay, Kiev, 93 pp. (In Russian).

Gilles G.L. 1968. Introduction to study of soil sickness in fruit cultures *Parasitica* **24** (1): 16-27. (In French).

Gilmore A.E. 1959. Growth of replanted peach trees. *Proc. Amer. Soc. Hortic. Sci.* **73**: 99-111.

Gilmore A.E. 1963. Pot experiments related to the peach replant problem. *Hilgardia* **34** (3): 63-78.

Golubynsky I. 1974. Repeated utilization of soil. *Tsvietovotstvo* **12**: 26. (In Russian).

Gortinsky G.B. 1963. Allelopathy and Biogeocenology. *Moscow Society of Nature Defenders. Section of Biology* **6**: 103-110. (In Russian).

Gortinsky G.B. 1966. Allelopathy in experiments of soviet scientists from the begining of XX century. *Moscow Society of Nature Defenders. Section of Biology* **21**: 71. (In Russian).

Grab T.A., Patyka W.F. 1976. Effect of some phytotoxic compounds produced by soil micro-organisms on plant cells. In: *Problems of Allelopathy*, Naukova Dumka, Kiev p. 17-18. (In Russian).

Greigh-Smith B. 1912. The bacteriotoxins and the agricere of soils. *Zbl. Bacteriol.* **11** (34): 224-227.

Grib N.I., Naydenko A.I., Postoyeva E. W. 1973. Effect of winter rye monoculture on its yield and fertility dark-grey podzol soil. *Agrochimia* **2**: 57-66. (In Russian).

Grieg-Smith R. 1918. Contribution to our knowledge of soil fertility. *Proc. Linn. Soc., New Wales* **63**: 169.

Grodzinsky A.M. 1962. Phytocenotic role of physiologically active secretions of plants (review). *Ukrain. Bot. J.* **29** (5): 3-20. (In Ukrainian).

Grodzinsky A.M. 1965. *Allelopathy in Life Plants and Their Surrounding.* Naukova Dumka, Kiev 198 pp. (In Russian).

Grodzinsky A.M. 1968. Determination of allelopathic properties of crop plants. *Proc. 3rd Conference of Physiologists and Biochemics of Siberia and Far East*. Edition of Sib. AN SSSR, Irkuck p. 38-40. (In Russian).

Grodzinsky A.M. 1969. Studies of allelopathy in agrophytocenoses. *Proc. 1st Workshop of Agrophytocenology*, Kazan p. 225-231. (In Russian).

Grodzinsky A.M. 1972. Mechanisms of action of volatile and water-soluble kolines. *Physiological and Biochemical Basis of Plant Interactions in Phytocenoses* 3: 3-7. (In Russian).

Grodzinsky A.M. 1973. *Basis of Chemical Interactions of Plants*. Naukova Dumka 205 pp. (In Ukrainian).

Grodzinsky A.M. 1974. Problems of soil sickness and allelopathy. *Physiological and Biochemical Basis of Plant Interactions in Phytocenoses* 5: 3-9. (In Russian).

Grodzinsky A.M. 1975. Cenotic studie in botanical gardens and their meaning in goal determination for defense of plant world. *Biol. Gl. Botan. Sada* 95: 23-28. (In Russian).

Grodzinsky A.M. 1976. Biogeocenology and introduction of plants. *Biol. Gl. Botan. Sada* **100**: 34-39. (In Russian).

Grodzinsky A.M. 1978. Treatment of seed with light lipophylic compounds. *Dop. AN USSR*, Ser. B 1: 50-53. (In Ukrainian).

Grodzinsky A.M., Bogdan G.P. 1972. Histochemical investigations of pectins, lignin and suberin in plants grown under influence of allelopathically active compounds. *Ukr. Bot. J.* **29** (2): 137-143. (In Ukrainian).

Grodzinsky A.M., Bogdan G.P. 1973. Role of ascorbic acid in formation of brown substances under influence of allelopathic factor. *Ukr. Bot. J.* **30** (1): 28-35. (In Ukrainian).

Grodzinsky A.M., Bogdan G.P. 1975. Effect of allelopathic factor on activity of ATP-ase in plants. *Physiological and Biochemical Basis of Plant Interactions in Phytocenoses* **6**: 15-19. (In Russian).

Grodzinsky A.M., Grodzinsky D.M. 1973. Germination method of Neubauer-Schneider. In: *Short Guideline of Plant Physiology*. Naukova Dumka, Kiev p. 506-507. (In Russian).

Grodzinsky A.M., Mohova N.I., Pilipienko-Yourtshak L.D., Filipovitsh T.M. 1962. Organic substances in post-harvest residiues and weeds. *Ukr. Bot. J.* 19 (6): 30-38. (In Ukrainian).

Grodzinsky A.M., Mohova N.I., Pilipienko-Yourtshak L.D., Filipovitsh T.M. 1963. organic substances in post-harvest residues and weeds. *Ukr. Bot. J.* **20** (1): 66-72. (In Ukrainian).

Grümmer G. 1957. *Relationships of Higher Plants. Allelopathy*. Izd. Innostr. Liter., Moscow, 261 pp. (In Russian).

Grümmer G. 1964. Role of toxic compounds in relationships of higher plants. In: *Mechanisms of Biochemical Competition*. Mir, Moscow, p. 277-288. (In Russian).

Grümmer G. 1970. Autotoxicity of flax. Physiological and Biochemical Basis of Plant Interactions in Phytocenoses 1: 49-55. (In Russian).

Guenzi W.D., Mc Calla T.M. 1966a. Phytotoxic substances extracted from soil. *Soil Sci. Amer. Proc.* **80** (2): 214-216.

Guenzi W.D., Mc Calla T.M. 1966b. Phenolic acids in oat, sorghum and corn residues and their phytotoxicity. *Agron. J.* **58** (2): 303-304.

Haziyev W.H. 1972. Soil enzymes and their role in fertility. *Biol. Nauki* **2**: 114-118. (In Russian).

Hein K. 1972. Contribution to soil sickness problem. *Gartenbauwissenschaft* **37** (1): 47-71. (In German).

Hennequin J.R., Juste C. 1967. Presence free phenolic acids in soil: Study of influence on germination and plant growth. *Ann. Agr.* **18** (5): 545-569. In French.

Hildebrand E.M. 1945. Peach root toxicity in New York orchard. *Plant Disease Rep.* **29** (6): 179.

Hoestra H., Oostenbrink M. 1962. Nematodes in relation to plant growth. *Pratylenchus penetrans* (Cobb.) on orchard trees. *Neth. J. Agr. Sci.* **10** (4): 74-77.

Hutchinson H., Thaysen A. 1918. The non persistence of bacterial toxins in the soil. *J. Agr. Sci.* **9**: 1.

Ivanov P.K. 1961. *Causes of Dying of Stone-fruit Trees*. Shtinca, Kishiniev 226 pp. (In Russian).

Ivanov S.M. 1973. Plant Extretions and Their Significance in Phytocenose Life. Nauka, Moscow 293 pp. (In Russian).

Ivanov W.P. 1974. Straw as manure. *Nautsh. Tr. Sarat. S.-H. Inst.* **12**: 294-334. (In Russian).

Janson A. 1936. *Seedy-, Stone- and Nut-trees*. Leipzig, 665 pp. (In German).

Jukova R.A. 1956. Microbiological studies of unploughed land soils of the Kola Peninsula. *Microbiologia* **25** (5): 569-576. (In Russian).

Jukova R.A. 1959. Distribution of aerobic cellulolytic bacteria in soils of the Kola Peninsula. *Microbiologia* **28** (1): 69-74. (In Russian).

Jukova R.A. 1964. Toxicity of the Kola Peninsula soils in relation to plants. *Potshvoviedienye*. **9**: 81-89. (In Russian).

Kalmykova N.A. 1975. Fungi-producers of phytotoxic substances as factors of chernozem soil toxicity. In: *Systematics, Ecology and Physiology of Soil Fungi*. Urotshay, Kiev p. 402-416. (In Russian).

Kaminsky R., Müller W.H. 1978. A recommendation against the use of alkaline soil extractions in the study of allelopathy. *Plant and Soil* **49** (1046): 641-645.

Kanivyetz I.I. 1970. General reasons of funtional diseases of crop plants and preventive measures. In: *Soil Conditions, Fertilization and Yielding of Crop and Fruit Cultures*. Urotshay, Kiev p. 402-416. (In Russian).

Kara U.M. 1976. Insecticidal plants. *Tsvietovotstvo* **7**: 20. (In Russian).

Katznelson R.S., Jershov W.W. 1968. Biological activity of Karel ASSR soils. *Microbiologia* **27** (1): 16-19. (In Russian).

Kefeli W.I. 1971. Relationships of phytohormones and natural inhibitors in plant growth. *Plant Physiology* **18** (3): 614-620. (In Russian).

Kimmel H.O. 1969. Activity of excretions of grass ornamental plants. *Proc. 4th Res. Scient. Conf. Young Researchers, Devoted 50-Anniversary of AN USSR*, Naukova Dumka, Kiev p. 79-81. (In Russian).

Kiselev G.E. 1952. *Perennial Ornamental Plants*. Selhozgiz, Moscow 972 pp. (In Russian).

Klaus H. 1940. Soil sickness problem in fruit farming. *Landwirt. Jahrb.* **89** (3): 413-459. (In German).

Klevenskaya I.L. 1974. *Oligonitrophilous Microorganisms of West Sybirian Soil*. Nauka, Novosibirsk, 219 pp. (In Russian).

Kobernuss E. 1951. Study of reasons and occurring soil sickness of fruit trees. *Kuhn-Arch.* **64**: 365-408. (In German).

Koch L.W. 1955. The peach replant problem in Ontario. I. Symptomatology and distribution. *Can. J. Bot.* **33** (4): 450-460.

Kossowitsh P.C. 1905. Clover soil sickness. *Journal Opyt. Agrochim.* **6**: 515. (In Russian).

Kotova I. 1974. Crop rotation system of grasslands. *Tsvietovotstvo* **8**: 5. (In Russian).

Kouprevitsh W.D. 1954. Intracellular enzymes in roots of higher autotrophic plants. *Dokl. AN SSSR* **18** (5): 953-956. (In Russian).

Kouprevitsh W.D. 1958. Problems of soil enzymology. *Vesti AN SSSR* **4**: 52-54. (In Russian).

Kouprevitsh W.D., Shtsherbakova T.A. 1966. *Soil Enzymology*. Nauka i Technika, Minsk 272 pp. (In Russian).

Kouranov W.N. 1961. Problems of plant residue decomposition. *Potshvoviedienye* **3**: 78-82.

Kovaltshouk J.G. 1973. Effect of endogenous ethylene on sprouts of peas, buckweed and barley. *Physiological and Biochemical Basis of Plant Interactions in Phytocenoses.* **4**: 65-67. (In Russian).

Kovaltshouk J.G., Boulkolova T.P. 1975. Effect of endogenous ethylene on some biochemical processes in peas sprouts. *Physiological and Biochemical Basis of Plant Interactions in Phytocenoses* **6**: 29-32. (In Russian).

Kovatshev G. 1970. Intensive cultures of fruit trees and problems of soil toxicity. *Ovoshtshevarstvo* **9**: 11-13. (In Bulgarian)

Kozel P.C., Tukey H.B. 1968. Loss of gibberellins by leaching from stems and foliage of *Chrysanthemum morifolium* 'Prince Anne'. *Amer. J. Bot.* **55** (10): 1184-1189.

Krall E.L. 1964. Nematodes – dangerous parasites. *Sadovotstvo* **4**: 50-51. (In Russian).

Kramer S., Schuricht R., Friedrich G. 1975. *Fruit Farming.* Deutsch. Landwirtschaftsverlag, Berlin. 245 pp. (In German).

Krasilnikov N.A. 1944. Phytohormonal activity of soil bacteria. *Dokl. AN SSSR* **45** (2): 87-90. (In Russian).

Krasilnikov N.A. 1958. *Soil Microorganisms and Higher Plants.* Izd. AN SSSR, Moscow 462 pp. (In Russian).

Krasilnikov N.A., Garkina N.R. 1946. Microbial factors of soil sickness. *Mikrobiologia* **15** (2): 109-114. (In Russian).

Laštůvka Z. 1961. Effect of active substances from quack grass on cell anatomy of wheat seedlings. In: *Katedra Fyziol. Rostlin, Brno* p. 373-392.

Lazar P., Chouroulinkow I., Libermann C., Guerin M. 1966. Benzo(a)pyrene carcinigenicity of cigarette smoke condensate – results of short-term and long-term tests. *J. Nat. Cancer Inst.* **37** (5): 573-581.

Lougauskas A.J., Shlaujene, D.J., Milkis R.I. 1976. Soil fungi in sick soils of clover fields. *Proc. Soviet Union Conf. "Problems and Methods of Biological Diagnostics and Estimation of Soils"*, Moscow p. 141-142. (In Russian).

Lucas E., Medicus F. 1886. *Science of Fruit Farming*, Stuttgart, 450 pp. (In German).

Manskaya S.M. 1957. Biosynthesis and decomposition of lignin. *Uspiehi Sovr. Biol.* **64** (1): 19-36. (In Russian).

Mario F. 1962. Succession of tree plants. *Inform. Hortofrutticolt.* **3**, N 19-30: 317-321. (In Italian).

Marle G.S. 1962. Soil sickness in cultures of fruit trees. *Meded. Direct. Tuinbouw* **25**: 114-122.

Martin J.P. 1948. Fungus flora of some California soils in relation to slow-decline of citrus tress. *Soil Sci. Amer. Proc.* **12** (3):209-214.

Martin J.P. 1948 a. Effect of fumigation, fertilization and various other soil treatments on growth of orange seedlings in old citrus soils. *Soil Sci.* **66** (3): 273-288.

Martin J.P. 1950. Effect of soil fungi on germination of sweet orange seeds and development of the young seedlings. *Soil Sci. Soc. Amer. Proc.* **14**: 184-188.

Martin J.P. 1950 a. Effect of fumigation and other soil treatments in the greenhouse on the fungus population of old citrus soil. *Soil Sci.* **69**: 107-122.

Martin J.P. 1950 b. Effect of various leaching treatments on growth of orange seedlings in old citrus soil. *Soil Sci.* **69**: 443-444.

Martin J.P. 1960. Fungi and nematodes in South African citrus orchard soils in relation to the citrus replant problem. *Soil Sci. Soc. Amer. Proc.* **24** (1): 91-94.

Martin J.P., Baines R.C., Erwin J.O. 1957. Influence of soil fumigation for citrus replants on the fungus population of the soil. *Soil Sci. Soc. Amer. Proc.* **21**: 163-166.

Martin J.P., Haider K. 1971. Microbiological activity in relation to soil humus formation. *Soil Sci.* **3** (1): 54-63.

Martin J.P., Pratt P.F. 1958. Fumigants, fungicides and the soil. *Agr. Food Chem.* **6**: 345-348.

McCalla T.M. 1971. Studies on phytotoxic substances from soil micro-organisms and crop residues at Lincoln, Nebraska. In: *Biochemical Interactions Among Plants.* Washington p. 39-44.

McPherson J.K., Müller C.H. 1969. Allelopathic effects of *Adestoma fasciculatum* "chamise", in California chaparral. *Ecol. Monogr.* **39** (2): 177-198.

Mehta N.C., Dubach P., Deuel H. 1961. Carbohydrates in the soil. *Adv. Carbohydrate Chem.* **16** (3): 335-355.

Meijneke C., Oostenbrink M. 1955. Soil sickness in nurseries and fruit-trees plantations. *Gartenbauwirtschaft* **3** (1): 31-32. (In German).

Midla H.I. 1965. Phloridzin as one of factors of soil sickness in orchards. *Agrochimia* **9**: 109-112. (In Russian).

Mirtshink T.G. 1957. Fungi responsible for toxicity of sward-podsolic soils with different degree of tillage. *Mikrobiologia* **26** (1): 74-86. (In Russian).

Mirtshink T.G. 1963. Distribution of fungi responsible for toxicity of different type soils and toxine forming in natural conditions. In: *Microorganisms in Agriculture*. Ed. Moscow University, Moscow p. 35-40. (In Russian).

Mirtshink T.G. 1972. Methods of estimation of fungi biomass in soils. In: *Problems of Number and Biomass of Soil Microorganisms*. Nauka, Lvov. (In Russian).

Mirtshink T.G. 1976. *Micology of soil*. Edition of Moscow University, 208 pp. (In Russian).

Mirtshink T.G., Stiepanova L.N. 1975. Dominant forms of fungi in sward-podsolic soils of Valday district. In: *Development Rules of Soil Micro-organisms*. Lvov, p. 242-248. (In Russian).

Mishustin E.N. 1956. *Microorganisms and Soil Fertility*. Izd. AN SSSR, Moscow 246 pp. (In Russian).

Mishustin E.N. 1961. Biological ways of increase of soil fertility efficiency. In: *Microorganisms and Fertility Efficiency of Soils*. Izd. AN SSSR, Moscow p. 121-133. (In Russian).

Mishustin E.N. 1975. *Associations of Soil Microorganisms*. Nauka, Moscow 107 pp. (In Russian).

Mishustin E.N., Naumova A.I. 1955. Formation of toxic compounds by alfalfa and their effect on cotton plants and soil microflora. *Izv. AN SSSR* Ser. Biologia 5: 3-9.

Mitsuchashi M., Shibaoka H. 1965. Isolation of an inhibitor of growth and root formation from *Portulaca grandiflora* leaves. *Plant and Cell Physiol.* 6 (1): 87-89.

Mohova N.I. 1962. Effect of volatile excretions on photosynthesis. *Proc. Sci. Confer. Experimental Botany*, Kazan p. 25-26. (In Russian).

Mohova N.I. 1966. Ecological role and action mechanism of miazmines. In: *Physiological and Biochemical Basis of Plant Interrelations in Phytocenoses*. Nauka, Moscow p. 167-173. (In Russian).

Molish H. 1937. *Influence One Plant on the Other: Allelopathy*. Iena Fisher 106 pp. (In German).

Moral R. del, Müller C.H. 1969. Fog drip: a mechanism of toxin transport from *Eucalyptus globulus*. *Bull. Torrey Bot. Club* 96 (4): 467-475.

Moral R. del, Müller C.H. 1970. The allelopathic effects of *Eucalyptus camaldulensis*. *Amer. Midland Natur.* 83 (1): 254-282.

Moroz P.A. 1968. *Allelopathic role of fallen leaves and root residues of apple and peach*. Autoref. Dis. Kand. Biol. Nauk. Kiev 25 pp. (In Russian).

Moroz P.A. 1972 a. Chemical character of saprolines, released from apple root residues. In: *Phytoncides: Results, Perspectives and Goals.* Naukova Dumka, Kiev p. 68-69. (In Russian).

Moroz P.A. 1972 b. Changes of toxicity of extracts from leaves nad roots of apple and peach in their relations with soil. *Physiological and Biochemical Basis of Plant Interactions in Phytocenoses* 3: 24-27. (In Russian).

Moroz P.A., Boukolova T.P. 1977. Phloridzin in soil from under apple. *Dop. AN USSR,* Ser. B, **5**: 447-449. (In Ukrainian).

Moroz P.A., Lashtshenko S.G., Artiemieva L.I. 1971. Effect of soil on toxic properties of extracts from weed residues. *Potshvoviedienye* **3**: 130-133. (In Russian).

Morozov A.W. 1976. Current problems of fruit farming. *Sadovotstvo* **4**:1-2. (In Russian).

Mouhin S.A. 1972. Superintensive orchard. *Sadovotstvo* **12**: 20-21. (In Russian).

Mountain W.B., Boyce H.R. 1958. The peach replant problem in Ontario. V. The relation of parasitic nematodes to regional differences in severity of peach replant failure. *Can. J. Bot.* 36 (1): 125-134.

Mountain W.B., Boyce H.R. 1958. The peach replant problem in Ontario. VI. The relation of *Pratylenchus penetrans* to the growth of young peach trees. *Can. J. Bot.* **36** (1): 135-151.

Mountain W.B., Patrick L.A. 1959. The peach replant problem in Ontario. VII. The pathogenity of *Pratylenchus penetrans. Can. J. Bot.* **37** (4): 459-470.

Mousyenko S.T., Mousyenko T.A. 1974. Monoculture of maize and its fertilization. *Agrochimia* **5**: 67-71. (In Russian).

Müller C.H. 1958. The association of desert animals with shrubs. *Amer. J. Bot.* **36** (2): 135-151.

Müller C.H. 1965. Inhibitory terpens volatilised from salvia shrubs. *Bull. Torrey Bot. Club* **92** (1): 38-45.

Müller C.H. 1966. The role of chemical inhibition (allelopathy) in vegetational composition. *Bull. Torrey Bot. Club* **93** (5): 219-221.

Müller C.H., Hanawalt, Mc Pherson. 1968. Allelopathic control of herb growth in the fire of California chaparral. *Bull Torrey Bot. Club* **95** (3): 225-231.

Müller C.H., Müller W.H. Haines B.L. 1964. Volatile growth inhibitors produced by shrubs. *Science* **143** (5): 471-473.

Müller K.H. 1970. New studies on allelopathy in America. In: *Physiological and Biochemical Basis of Plant Interrelations in Phytocenose.* Nauka, Moscow p. 167-173. (In Russian).

Müller W.H., Muller C.H. 1964. Volatile growth inhibitors produced by *Salvia* species. *Bull. Torrey Bot. Club* **91** (4): 171-174.

Mylrs B.D., Throneberry G.O. 1966. Effect of gossypol on some oxidative respiratory enzymes. *Plant Physiol.* **41** (5): 787-791.

Nadkernitshny S.P. 1975. Phytotoxic compounds and species complex of microscopic fungi in strong, weak-humus chernozem in crop rotation and monoculture system of winter wheat. In: *Systematics of Ecology and Physiology of Soil Fungi.* Naukova Dumka, Kiev p. 72-73. (In Russian).

Nagovicina A.W., Nikitina K.B. 1975. Pecularities of development of root rot on alfalfa and clover plants in non-chernozem area. *Proc. 6th Soviet Union Conf. "Resistance Agriculture Plants on Diseases and Pests"*, Moscow p. 269-270. (In Russian).

Narcisov W.P. 1976. *Scientific Basis of Tillage Systems.* Kolos, Moscow 354 pp. (In Russian).

Naumova A.N. 1959. Accumulation of toxic compounds in grey soil under alfalfa and their effect on microflora. In: *Microbiology serving the agriculture.* Selhozgiz, Moscow p. 39-43. (In Russian).

Negre R. 1968. Lithospermum fructicosum and Rosmarinus officinalis. Phyton **13** (1/2): 79-83. In French.

Nikitina K.W., Nagovitsina A.W. 1972. Root root of perrennial papilionaceous plants on non-chernozem area and role of bacteria in this disease. *Proc. Confer. on Bacterial Plant Diseases,* Kiev p. 58-60. (In Russian).

Norstadt F.A., McCalla T.M. 1968. Microbially induced phytotoxicity in stublemulched soil. *Proc. Soil Sci. Soc. Amer.* **32** (2):241-245.

Norstadt F.A., McCalla T.M. 1969. Patulin production by *Penicillium urticae* Bainer in batch culture. *Appl. Microbiol.* 17 (2): 193-196.

Norstadt F.A., McCalla T.M. 1971. Effects of patulin on wheat grown to maturity. *Soil Sci.* **3** (4): 236-243.

Ohman J.H. Kommedahl Th. 1960. Relative toxicity of extracts from vegetative organs of quackgrass to alfalfa. *Weeds* 8 (4): 666-670.

Oleksyevitsh W.M. 1971. *Learning of Allelopathic Properties of Some Tree and Bush Plants Used in Greenery.* Autoref. Dis. Kand. Biol. Nauk. Dnietropetrovsk 25 pp. (In Russian).

Oryel S.E., Boyko P.I. 1975. *Correct Crop Rotation – Guarantee of High Yield.* Urotshay, Kiev p. 13-20. (In Ukrainian).

Otto G. 1966. Soil sickness – specific decrease of soil fertility. *Obstbau* **6** (2): 19-20. (In German).

Papadakis 1941. *Soil Sci.* **52** (2): 283-290.

Parenti R.L., Rice E.L. 1969. Inhibitional effects of *Digitaria sanguinalis* and possible role in oldfield succession. *Bull. Torrey Bot. Club* **96** (1): 70-78.

Parsons J.W., Tinsley J. 1960. Extraction of soil organic matter with anhydrous formic acid. *Soil Sci. Am. Proc.* **24**: 198-201.

Parsons J.W., Tinsley J. 1961. Chemical studies of polysaccharide material in soils and composts based on extraction with anhydrous formic acid. *Soil Sci.* **92** (1): 46-53.

Patrick Z. 1955. The peach replant problem in Ontario. II. Toxic substances from microbiological decomposition products peach root residues. *Can J. Bot.* **33** (4): 461-486.

Pengerot J. 1913. What way to renowate old died orchards? *Sel. Hozyain.* **25**:1313-1315. (In Russian).

Persidsky D.I., Wilde S.A. 1954. The effect of volatile substances released by soil, humus and composts on the growth of excised roots. *Plant Physiol.* **29** (5):484-486.

Pickering S. 1903. The effect of one plant upon another. *J. Roy. Agr. Soc.* **64**: 365-376.

Pickering S. 1910. Action of one crop on another. *J. Roy. Hort. Soc.* **43**: 372-380.

Pickering S. 1917. The effect of one plant on another. *Ann. Bot.* **31** (2): 181-187.

Pieniazek F.T. 1973. *When Will Apple Trees Be Blooming?* Mir, Moscow 464 pp. (In Russian).

Pitcher R.S., Way D.W., Savory B.M. 1966. Specific replant diseases of apple and cherry and their control by soil fumigation. *Hort-Science* **41** (4): 377-396.

Plenk J. 1975. *Physiology and Pathology of Plant*, Vien. (In German).

Plhak F. 1972. Effect of volatile excretions of osier willow leaves on oxidative phosphorylation in rye sprouts. *Physiological and Biochemical Basis of Plant Interactions in Phytocenoses* **3**: 16-21. (In Russian).

Polishtshuk O.D., Snishko W.L. 1954. *Renovation of Old Orchards*. Kiev 11 pp. (In Ukrainian).

Ponomaryeva W.W. 1947. Methods of fulvic acid isolations and their chemical character. *Potshvoviedienie* **12**: 714-723. (In Russian).

Popivshtshy I.I. 1972. Biological activity of water extracts from roots of wild fruit plants. In: *Basis of Chemical Interrelations of Plants in Phytocenoses.* Naukova Dumka, Kiev p. 82-84. (In Russian).

Popov W.N. 1974. Intensification – the general direction. *Sadovotstvo* **3**: 1-6. (In Russian).

Popova E. 1973. Onion seeds and aster. *Tsvetovotstvo* **7**: 30. (In Russian).

Poptzov A.W. 1959. Using of biological method for estimation of usefulness soil mixtures for plant growing. *Biol. Gl. Botan. Sada* **35**: 107-109. (In Russian).

Posner A.M. 1966. The humic acid extracted by various reagents from a soil. Part I. Yield, inorganic components and titration curves. *J. Soil Sci.* **17** (1): 65-78.

Pozrebnyak P.C. 1968. *General Forestry*. Kolos, Moscow 291 pp. (In Russian).

Proebsting E.A. 1950. A case history of a peach replant situation. *Proc. Amer. Soc. Hort. Sci.* **59**: 319-321.

Proebsting E.A., Gilmore A.E. 1940. The relation of peach root toxicity to the reestablishing of peach orchards. *Proc. Amer. Soc. Hort. Sci.* **38**: 21-26.

Proutenskaya N.I. 1975. Allelopathic soil sickness uder flower cultures. *Physiological and Biochemical Basis of Plant Interactions in Phytocenoses*. **6**: 61-66. (In Russian).

Pryanishnikov D.N. 1976. Crop Rotation and Its Significance in Works of Learning Our Crops. Nauka, Moscow p. 372-403. (In Russian).

Pyerov N.N., Mirzayev M.N. 1971. Role of microflora in grapevine toxicosis infested *Phylloxera*. *Dokl. WASHIL* **9**: 24. (In Russian).

Pyetrenko I.A. 1973. *Annual Asters*. Lenizdat., Leningrad 136 pp. (In Russian).

Pyetrenko N.B., Styepanenko A.J., Safonova I.P., Kozlovskaya L.G. 1971. To methodology learning soil toxicity nature under sugar beet. In: *Microbiological and Biochemical Studies of Soils*. Urotshay, Kiev p. 222-226. (In Russian).

Rademacher B. 1959. Interrelations of higher plants. In: *Guideline of Plant Physiology*, Springer, Berlin, p. 655-706. (In German).

Rafaelyan Z.C. 1970. Dynamics of available trophic compounds in grey soil and yield of winter wheat in monoculture and systematic application of fertilizers. *Agrochimia* **8**: 72-78. (In Russian).

Rauhe K., Lehne I. 1966. Conclusions to results of continuous experiment on fertilization. "Eternal rye culture" in vegetation house in view of fertility of soil. *Albrecht Thaer Arch.* **10** (1): 190-193. (In German).

Reber M., Schara A. 1971. Degradation sequences in wheat straw extracts inoculated with soil expansions. *Soil Biol. and Biochem.* **3** (4): 381-383.

Reitner E.I., Smirnov A.M., Chun-Chou H. Uhina C.F., Kuzovkina I.N., Roizin M.B., Romeyko I.K. 1963. Amino acids availability in presence of nitrogen source by isolated alfalfa roots and intact peas plants in sterile culture. *Plant Physiology* **10** (6): 673-681. (In Russian).

Rice E.L. 1965. Inhibition of nitrogen-fixing and nitrifying bacteria by seed plants. III. Comparison of three species of *Euphorbia*. *Proc. Okla. Acad. Sci.* **45** (1): 43-44.

Rice E.L. 1974. *Allelopathy*. New York, Academic Press. 353 pp.

Rice E.L. 1978. *Allelopathy*. Mir, 391 pp. (In Russian).

Roizin M.B. 1969. Seasonal changes of podsolic soil toxicity. In: *Soil Regimes on the Polar North*. Nauka, Lvov p. 132-138. (In Russian).

Romeyko I.M., Bityukova L.B. Zil L.B., Koutsherenko W.I. 1976. Comparative characteristics of methods showing phytotoxic properties of soil microorganisms. *Mikrobiol. J.* **38** (4): 456-461. (In Ukrainian).

Romeyko I.M.1969. Proteolytic activity of sward-podsolic soil with different methods of ploughing. *Potshvoviedienie* **10**: 87-91. (In Russian).

Roshtshina W.D. 1975. Some physico-chemical and biological methods of investigation of plant excretions. *Physiological and Biochemical Basis of Plant Interactions in Phytocenoses* **6**: 109-114. (In Russian).

Roshtshina W.W. Akulova E.A. 1976. Effect of phloridzin on phytosynthetic transport on cytochrome level. In: *Problems of Allelopathy*. Naukova Dumka, Kiev p. 32-33. (In Russian).

Roskoshansky A.B. 1976. Post-harvest root residues and their role in changes of soil fertility conditions. *Agrochimia* **1**: 95-99. (In Russian).

Russel E.D. 1936. *Soil Conditions and Plant Growth*. Selhozgiz. 501 pp. (In Russian).

Russel E.D. 1955. *Soil Conditions and Plant Growth*. Izd. Inostr. Lit. 624 pp. (In Russian).

Sahaboutdinova A.Z. 1975. Effect of crop rotation and monoculture on microflora and some biochemical processes in leached chernozems. In: *Bashkirian Soils*, Ufa p. 204-216. (In Russian).

Salayev R.K. 1969. *Sorption of Substances by Plant Cells*. Nauka, Moscow 186 pp. (In Russian).

Samtsevitsh S.A., Borisova W.N. 1963. Toxicity of volatile substances formed by microorganisms in soil. *Mikrobiologia* **32** (3): 484-491. (In Russian).

Samtsevitsh S.A. 1966. Gel-like excretions of plant roots and their effect on microorganisms. In: *Physiological and Biochemical Basis of Plant Interrelations in Phytocenoses*. Nauka, Moscow p. 254-261. (In Russian).

Samtsevitsh S.A., Romanov S.L. 1975. Chemical composition of gel-like compounds, produced by maize apical points. *Wiesti AN BSSR* Ser. Bil. Navuk. **6**: 105-106. (In Russian).

Schander H. 1956. *Soil Sickness in Fruit-tree Cultures*. Bonn, 66 pp. (In German).

Schnitzer M., Wright I.K. 1956. Note on the extraction of organic matter from the B horizon of podzol soil. *Can. J. Agr. Sci.* **36** (5): 511-512.

Schreiner O., Reed H. 1907. Some factors influencing soil fertility. *U.S. Dept. Agr., Bur. Soil Bull.* **40**: 12-15.

Schreiner O., Reed H. 1908. The toxic action of certain organic plant constituents. *Bot. Gaz.* **45**: 73-102.

Schwär C. 1962. Effect of extracts from wormwood on *Foeniculum vulgare* Mill., *Lepidium sativum* L. and *Lactuca sativa* L. var *longifolia* Lam. *Flora* **152** (3): 509-515. (In German).

Shkolnaya Z.P. 1976. Jonquils in subtropical region. *Tsvietovotstvo* **10**: 2. (In Russian).

Shmygun W.N. 1972. *Chrysanthemums*. Nauka, Moscow 114 pp. (In Russian).

Shtsherbanovsky L.P., Shubina L.S., Blagonravova L.N. 1976. Effect of kolines on enzymatic systems acceptor-plants and microorganisms. In: *Problems of Allelopathy*. Naukova Dumka, Kiev p. 41-42. (In Russian).

Shtshurova G.W. 1976. For growing of healthy tulips. *Tsvietovotstvo* **8**: 20. (In Russian).

Simakov W.N., Ceplyenkov W.P. 1970. Modification of volume method of humus determination (for mass analyses). *Potshvoviedienye* **4**: 113-119. (In Russian).

Smirnova W.W. 1969. On Problem of Influence of Biocenotic Environment on Resistance of Gladiolus on Vessel Fusariose. Ed. Kazan University, Kazanp. p.141-146. (In Russian).

Souhovyetsky A.I. 1972. Agroclimatic characteristic of region of breeding Mironovsk cultivars of winter wheat. In: *Mironovsk Wheats*. Kolos, Moscow p. 13-20. (In Russian).

Sourovaya T.P., Goryelik I.I. 1974. Sickness of Artificial Soil under Influence of Monoculture and Possible Role of Microorganisms in This Procces. Izd. AN BSSR. Ser. B, **5**: 53-58. (In Russian).

Sovyetov A.B. 1950. *On Tillage Systems*. Gosselhozizdat., Moscow p. 235-419. (In Russian).

Stom D.I. 1975. Allelopathy and hypothesis on quinones as active form polyphenols *Physiological and Biochemical Basis of Plant Interactions in Phytocenoses* **6**: 8-11. (In Russian).

Styepanov S.N. 1976. Problems of fruit farming development. *Sadovotstvo* **6**: 34. (In Russian).

Styepanova L.P. 1971. Effect of inhibitors from alfalfa 'AzSHI-1' and AzNIHI-262 on growth and development of next cultures. *Izd. AH AzSSR*. Ser. Biol. Nauk. 1: 43-48. (In Russian).

Svishtshuk A.A., Sereduk L.S., Moroz P.A. 1977. fatty acids in soil under from apple. In: *Interrelations of Plants and Microorganisms in Phytocenosis*. Naukova Dumka, Kiev p. 65-69. (In Russian).

Svishtshuk A.A., Sereduk L.S., Moroz P.A., Shevtshek T.N. 1975. Methods of isolation of organic compounds from soils. *Physiological and Biochemical Basis of Plant Interactions in Phytocenoses* 6: 115-117. (In Russian).

Swart-Füchtbauer H. 1954. Ectoparasitic nematodes as probable reason of soil sickness in nursery. *Naturwissenschaften* **41** (2):148-149. (In German).

Swincer G.D., Oades J.M., Greenland D.J. 1969. The extraction, characterization and significance of soil polysaccharides. *Adv. Agron.* **21** (2): 195-235.

Syrovatskaya L. 1974. rational crop rotation system. *Tsvietovotstvo* 1: 12. (In Russian).

Takijima Y. 1964a. Studies on organic acids in paddy field soils with reference to their inhibitory effects on the growth of rice plants. *J. Soil Sci. Plant Nutr.* 10 (1): 22-29.

Tamura S. 1969. Chemical studies on "clover sickness". Part I. Isolation and structural elucidation of two isoflavonoides into red clover. *Agr. Biol. Chem.* **33** (3): 391-397.

Tarasov W.N. 1966. Fight with apple rosette disease. *Sadovotstvo* **3**: 16-18. (In Russian).

Tarasov W.N. 1972. *Apple Rosette Disease*. Urotshay, Kiev 176 pp. (In Russian).

Tatarinov P.N., Pavlov G.D. 1976. *Fruit Farming on Weak-growing Rootstocks*. Urotshay, Kiev 176 pp. (In Russian).

Terentyev W.N., Tsaryeva W.I., Shoutskaya O.B. 1969. Method of aldehyde determination in soil. In: *Photosynthesis and Plant Nutrition*. Nauka i Technika, Minsk p. 56-63. (In Russian).

Timiryazev K.A. 1949. *Selected Works*. Selhozgiz, Moscow vol. 3: 117 pp. (In Russian).

Tinsley J., Salam A. 1961. Extraction of soil organic matter with aqueous solvents. *Soils and Fert.* **24** (1): 81-84.

Tokin B.P. 1974. *Remedial Toxins of Plants*. Lenizdat, Leningrad 213 pp. (In Russian).

Treibs W. 1947. On agropyren – natural aromatic hydrocarbon in roots of quack grass. *Chem. Ber.* **80** (1): 97-101. (In German).

Treibs W. 1959. Agropyren from aromatic oil of quack grass roots. Phenylhexadien with wide antibiotic effect. *Parfüm. und Kosmet.* **40** (5):451-452. (In German).

Tsaryeva R.I., Semenova T.A., 1971. Identification and quantitive determination of phenolocarboxylic acids in peat soil. In: *Plastide Apparatus and Vital Activity of Plants.* Nauka i Technika, Minsk p. 95-100. (In Russian).

Tshasovyennaya A.A. 1972. Significance of biologically active substances releasing into envirionment for plants of different developmental stage. *Proc. 3rd Soviet Union Symp. "Basis of Chemical Interrelations in Phytocenoses"* Naukova Dumka, Kiev p. 54-55. (In Russian).

Tshernomaz P.A. Korolkov L.P. 1963. Experiment on maize monoculture in Tambovsk region. *Dokl. Timiriazev. S.-H. Akad. Rastienievodstvo* **93**: 5-12. (In Russian).

Tshunderova A.I. 1964. Infection of millet root system by *Fusarium oxysporum* and *Helminthosporum* sp. – general reason millet soil sickness. *Mikrobiologia* **33** (1): 292-298. (In Russian).

Tshunderova A.I. 1968. Effect of Carbohydrazide in Sward-podsolic Soils with Different Content of Humus. Nauka i Technika, Minsk p. 172-179. (In Russian).

Tshunderova A.I., Shoulyndin A.F. 1964. Millet soil sickness and their reasons. *Zemledelie* **2**: 84-88. (In Russian).

Tshunderova A.I., Zubetz T.P. 1968. Effect of monoculture on intensity of biochemical processes sward-podsolic soil. *Proc. Symp. "Enzymes of Soil"*, Nauka i Technika, Minsk p. 164-171. (In Russian).

Tukey H.B. Jr. 1966. Leaching of metabolites from above- ground plant parts and its implications. *Bull. Torrey Club* **93** (6): 385-401.

Tukey H.B. Jr. 1969. Implications of allelopathy in agricultural plant science. *Bot. Rev.* **35** (1): 1-16.

Ugrehelidze D.Sh. 1974. Effect of simple phenols on auxin metabolism in maize roots. *Physiological and Biochemical Basis of Plant Interactions in Phytocenoses* **5**: 44-46. (In Russian).

Uhtomskaya F.I. 1952. Role of enzymes in soil auto-purifying. *Gigiena i Sanitaria.* **11**: 46-54. (In Russian).

Valadand D., Cezard C. 1971. Study on biological activity of hydroquinone: Effect on cell divisions. *C.R. Acad. Sci. D.* **273** (23): 2247-2248.

Vaverkova-Cambalikova S. 1971. Cytological studies on effect of tetracyclic antibiotics on plant meristems. *Biologen* (CSSR) **26** (10): 777-780.

Vergilius Publius Maron. 1971. *Boucolics. Georgicon. Eneida.* Hudoshestviennaya Literatura. 442 pp. (In Russian).

Verona P.L. 1963. Soil sickness. *Agr. Ital.* **63** (18): 283-335. (In Italian).

Vicherkova M. 1970. Effect of active compounds from rhizomes of quack-grass on growth and water relationships of flax. *Physiological and Biochemical Basis of Plant Interactions in Phytocenoses* **1**: 135-141. (In Russian).

Vinogradsky S.N. 1951. *Soil Microbiology*. Edition of AN USSR, Moscow. (In Russian).

Vogel F., Weber E. 1931. Soil sickness in nurseries of fruit trees. *Gartenbauwissenschaften* **5**: 508-524. (In German).

Vorobyev C.A. Intensification of tillage and crop rotation. *Izviestija Timiriazev Sielsko-Hozaistviennoy Academy* **6**: 16-27. (In Russian).

Voynova G. Dzumalyeva D., Hlyebarova C. 1975. Influence of culture reduction in crop rotation system on biological processes in soil. *Proc. 3rd Congress of Microbiology*, Sofia p. 243-248. (In Russian).

Walters E.H. 1971. The isolation of p-hydroxybenzoic acid from soil. *Amer. Chem. Soc. J.* **39** (2): 1778-1791.

Wang T.S.C., Cheng S.-Y., Tung H. 1967. Dynamics of soil organic acids. *Soil Sci.* **104** (2): 138-144.

Wang T.S.C., Cheng S.-Y., Tung H. 1967. Extraction and analysis of soil organic acids. *Soil Sci.* **103** (5): 360-366.

Wang T.S.C., Chuang T.T. 1967. Soil alcohols, their dynamics and their effect upon plant growth. *Soil Sci.* **104** (1): 40-45.

Wensley R.N. 1956. The peach replant problem in Ontario. IV. Fungi associated with replant failure and their importance in fumigated and nonfumigated soils. *Can. J. Bot.* **34** (6): 967-981.

Whitehead D.C. 1964. Identification of hydrobenzoic, vanillic, p-coumaric and ferulic acids in soils. *Nature* **202**: 417.

Whitehead D.C., Tinsley J. 1964. Extraction of soil organic matter with dimethylformamide. *Soil Sci.* **97** (1): 34-42.

Whitney M., Cameron F. 1903. The chemistry of the soil as related to crop production. *U.S. Dept. Agr. Bur. Soils Bull.* **22** (4): 413-433.

Whittaker R.H. 1971. The chemistry of communities. In: *Biochemical Interactions Among Plants*, Washington, Nat. Acad. Sci.: 10-15.

Winter A., Street H.E. 1963. A new natural auxin isolated from "staled" root culture medium. *Nature* **198** (4887): 1283-1288.

Winter A.G. 1955. Study on occurring and importance of antimicrobial and phytotoxic substances in natural soils. *Z. Pflanzenernahr. Bodenkd.* **69** (1/3): 224-233. (In German).

Winter A.G. 1960. Fertility and soil – input to problem of active substance cycling. *Munch. Mediz. Wochenzeitschrift* **102**: 2433-2438. (In German).

Yaryeshko W. 1973. Marigold – insecticide. *Tsvietovotstvo* **5**: 28. (In Russian).

Yegorov W.E. 1958. Results of long-lasting experiment on fertilization, crop rotation systems and monocultures. *Izv. Timiriazev. S.-H. Akad.* **3**: 29-58. (In Russian).

Yegorov W.E. 1972. *60 Years Lasting Experiment*. Znanie, Moscow, 48 pp. (In Russian).

Yermakov A.I., Arasimovitsh W.W., Smirnova M.I., Ikonnikova M.I., Yarosh N.P., G.A. Lukovnikova. 1952. *Methods of Biochemical Studying of Plants*. Kolos, Moscow 520 pp. (In Russian).

Yevdokimova G.A., Berednikova T.A. 1976. Microbial activators and inhibitors in rhizosphere of cereal grasses. In: *Soil Studies on the Kola Peninsula: Apatites*. p. 90-97. (In Russian).

Yourtshak L.D. 1971. *Physiologically Active Compounds of Common Serradella and Accompanying Microflora*. Autorefer. Dis. Kand. Biol Nauk. Kiev, 25 pp. (In Russian).

Yourin P.W., Shef R.P., Tshernysheva W.I. 1972. Role of free radicals in phsiological and biochemical interactions of plats in phytocenoses. *Physiological and Biochemical Basis of Plant Interactions in Phytocenoses* **3**: 127-134. (In Russian).

Zabolotskaya T.G., Lutoyeva M.H. 1974. Post-harvest residues of crop cultures, their decomposition and effect on holding available forms of nitrogen in some podsolic soils. *Agrochimia* **2**: 3-7. (In Russian).

Zahartshenko I.T., Pirozenko G.S., Suhobrous S.W. 1962. Effect of monoculture on fertility of soil. *Potshvoviedienye* **7**: 10-17. (In Russian).

Zaitshenko O.M. 1976. Soviet Union Scientific and Methodological Conference "Phytotoxicity of soil microorganisms", Tshernihov. *Microbiol. J.* **38** (6): 792-793. (In Russian).

Zalyalov F.K. 1967. Monocultures of some crops with different systems of fertilization. *Dokl. Timiriazev. S.-H. Akad.* **131**: 163-169. (In Russian).

Zavarzin G.A. 1972. *Methotrophic Microorganisms*. Nauka, Moscow, 192 pp. (In Russian).

Zubyenko W.F. 1973. Balance of organic compounds in different crop rotation systems sward-podsolic soils. *Agrochimia* **4**: 61-69. (In Russian).

Zviagintsev D.G. 1975. Some problems of microorganism ecology. *Proc. of Soviet Union Conference of Microbiological Society*, Erevan p. 1-2. (In Russian).

APPENDIX

1. Names of plants species used in the Book

Common Name	English Name	Botanical Name
Trees and shrubs		
Apple	Apple-tree	*Malus domestica* Borkh.
Apricot	Apricot	*Armeniaca vulgaris* Lam.
Banana	Common banana	*Musa x sapientum* L.
Beech	European beech	*Fagus Sivatica* L.
Blackthorn	Blackthorn	*Prunus Spinosa L.*
Cherry	Sour cherry	Prunus cerasus L
Currant	Black currant	*Ribes nigrum* L.
Gooseberry	English gooseberry	*Ribes grossularia* L.
Grape	Vine grape	*Vitis vinifera* L
Pea tree	Siberian pea tree	*Caragana arborescens* Lam.
Peach	Peach	*Persica vulgaris* Mill.
Pear	European pear	*Pyrus communis* L.
Pine	Common pine	*Pinus sylvestris* L.
Pineapple	Pine apple	*Ananas comosus* (L.) Merill.
Plane-tree	American plane	*Platanus occidentalis* L.
Plum	Garden plum	*Prunus domesticca* L.
Quince	Quince	*Cydoia oblonga* Milla.
Raspberry	Red raspberry	*Rubus idaeus* L.
Spruce	Common spruce	*Picea abies* (L.) H. Karst.
Sweet cherry	Sweet cherry	*Prunus avium* L.

Walnut	Black sea walnut	*Juglans regia* L.
Crop plants		
Alfalfa	Blue Lucerne	*Medicago sativa* L.
Barley	Barley	*Hordeum vulgare* L.
Batat	Sweat potato	*Ipomea batatas* (L.) Lam.
Bean	Common bean	*Phaseolus vulgaris* L.
Beet	Sugar beet	*Beta vulgaris* L.
Black mustard	Black mustard	*Brassica nigra* (L.) Koch.
Black radish	Spanish radish	*Raphanus sativus* L. var. *niger*
Buckwheat	Beech wheat	*Fagopyrum esculentum* Moench
Chick pea	Indian gram	*Cicer arietinum* L.
Clover	Red clover	*Trifolium pratense* L.
Cress	Garden cress	*Lepidium sativum* L.
Cucumber	Cucumber	*Cucumis sativus* L.
Flax	Common flax	*Linum usitatissimum* L.
Italian ryegrass	Italian rye-grass	*Lolium multiflorum* L.
Maize	Corn	*Zea mays* L.
Marjorma	Marjoram	*Origanum majorana* L.
Millet	True millet	*Panicum milllaceum* L.
Oat	Common oat	*Avena sativa* L.
Onion	Onion	*Allium cepa* L.
Pea	Field pea	*Pisum sativum* L.
Potato	Potato	*Sulanum tuberosum* L.
Radish	Turnip radish	*Raphanus sativus* var. *radicula* Pers.
Rape	Turnip	*Brassica rapa* L.
Rye	Rye	*Secale cereale* L.
Strawberry	Garden strawberry	*Fragaria ananassa* Duchesne
Sugar beet	Sugar beet	*Beta vulgaris* var. *altissima* DC
Sugar cane	Sugar cane	*Saccharum officinarum* L.
Sunflower	Common sunflower	*Helianthus annuus* L.
Tobacco	Virginian tobacco	*Nicotiana tabacum* L.
Tomato	Tomato	*Lycopesicon esculentum* Mill.

Wheat	Bread wheat	*Triticum aestivum* L.
White mustard	White mustard	*Sinapsis alba* L.
Ornamental plants		
Alexandrian senna	Tinnevelly senna	*Senna alexandrina* Mill.
Aster	China-aster	*Callistephus chinensis* Nees.
Balsam	Garden balsamine	*Impatiens balsamina* L.
Begonia	Perpetual begonia	*Begonia semperflorens* Link et Otto.
Chryzanthemum	Mother chrysanthemum	*Chrysanthemum japonicum* Thunb.
Delphinium	Florists larkspur	*Delphinium x cultorum* Voss
Dicentra	Showy dicentra	*Dicentra spectabilis* L.
Flox	Summer flox	*Phlox paniculata* L.
Gerbera	Flameray gerbera	*Gerbera jamesonii* Bolus.
Gladiolus	Gandavian cornflag	*Gladiolus x hortulans* L.H. Bailey
Heather	Common heather	*Calluna vulgaris* Salisb.
Hyacinth	Dutch hiacinth	*Hyacinthus orientalis* L.
Lupine	Tarwi lupin	*Lupinus mutabilis* Sweet
Narcissus	Pheasant's eye	*Narcissus poeticus* L.
Pink	Carnation	*Dianthus caryophyllus* L.
Poppy	Garden poppy	*Papaver somniferum* L.
Sweet pea	Sweet pea	*Lathyrus odoratus* L.
Tulip	Garden tulip	*Trlipa gesneriana* L.
Violet	Sweet violet	*Viola ododrata* L.

SUBJECT INDEX

AUTHOR INDEX